W9-DHK-380

WITHDRAWN

# SEQUENCES, COMBINATIONS, LIMITS

Library of School Mathematics

I. M. Gelfand, *General Editor*

American editions prepared by the Survey of Recent East European Mathematical Literature at the University of Chicago

**Volume 1: The Method of Coordinates**
by I. M. Gelfand, E. G. Glagoleva, and A. A. Kirillov

**Volume 2: Functions and Graphs**
by I. M. Gelfand, E. G. Glagoleva, and E. E. Shnol

Library of School Mathematics

Volume 3

# SEQUENCES, COMBINATIONS, LIMITS

S. I. Gelfand, M. L. Gerver, A. A. Kirillov,
N. N. Konstantinov, and A. G. Kushnirenko

*Translated and adapted from the Russian by*
Leslie Cohn *and* Joan Teller

*SURVEY OF*
*RECENT EAST EUROPEAN*
*MATHEMATICAL LITERATURE*

*A project conducted by*

Izaak Wirszup
*Department of Mathematics,*
*The University of Chicago,*
*under a grant from the National Science Foundation*

THE M.I.T. PRESS
Massachusetts Institute of Technology
Cambridge, Massachusetts, and London, England

Library of Congress Catalog Card Number: 68-14454
Printed in the United States of America

## Foreword

This volume, the third in the series "The Library of School Mathematics," is a collection of problems for the ninth and tenth grades.

This collection has been designed in such a way that it can also serve as a textbook for parts of a course.

Our presentation differs in many ways from the traditional.

In the first place, greater attention is given to the presentation of theory than to computations.

In the second place, we rely heavily on the independent work of those who study this book. The third essential difference is in the choice of material. We have tried to teach that part of high school mathematics which will prove particularly useful for future studies and work.

This book can be read in various ways. The simplest and most useful way is to read the conditions of the problems, read and analyze the solutions, and then solve the test problems. We think that the student can master the material by this method, passive though it may be. But it is somewhat more efficient to proceed as follows: Without looking at the answers, hints, and solutions, solve the entire series of problems independently. You can then compare your results with the answers that have been provided. Then read the solutions. We recommend that this be done even if the problems do not involve any special difficulties, since in the solutions we frequently give useful supplementary information, and sometimes we raise new questions.

If this method is too difficult for you, you can take an intermediate path: If you have been unable to solve certain problems even after a number of attempts, look at the section entitled "Answers and Hints." If the problem does not come out even after this, read the solution.

We shall make use of "road signs," as in Volume 1.

The "Parking Permitted" sign marks the passages containing information necessary for understanding the following: definitions and formulas, and so on. At this sign one should stop, read the passage carefully.

The "Steep Going" sign is placed at those passages containing more difficult material. If the passage is in small script, it may be omitted in the first reading.

Pay particular attention to the "Dangerous Turn" sign. Frequently we have placed it at those passages where at first glance everything seems easy. However, if the passage is not analyzed properly, serious errors may arise in what follows.

Usually a problem book serves as a supplement to a textbook. When working with our booklet, however, one cannot use a textbook.

This method of treating the material has been practiced in several of the mathematics schools of Moscow since the school year of 1962–1963.[1]

Two more remarks:

1. Most probably, you will find combinations the

[1]See the article by M. L. Gerver, N. N. Konstantinov, and A. G. Kushnirenko, "Zadachi po algebre i analizu" (Problems in Algebra and Analysis) in the journal *Obuchenie v matematicheskikh shkolakh* (Instruction in Schools of Mathematics), Volume 1, Moscow, "Prosveshchenie" (Education), 1965. Part of the material from this article has been incorporated in this book.

most interesting topic to start with. You need not be afraid to open the book at page 14 and begin to solve the problems of Chapter 2 immediately (Problems 44, 46, and 50!). To solve problem 44, for example, you do not need to know even algebra, and so you could give this problem (or the problem about the rattles) to your younger brother or sister.

2. This booklet contains many problems in which it is necessary to prove an assertion for all $n$. Do not let this frighten you! Check the assertion first for $n = 2, 3$, and 4; in doing this you will become aware of a number of regularities that will help you (and this is most important) to get a feeling for the conditions of the problem, and sometimes they will even suggest the method of solution (see, for example, Problem 2).

Chapter 1 was written by S. I. Gelfand; Chapter 2, by M. L. Gerver and A. G. Kushnirenko; and Chapter 3, by A. A. Kirillov. N. N. Konstantinov participated in the selection of the problems and in the editing of each of the chapters.

The authors wish to express their thanks to a number of colleagues who have helped in the preparation of this book: I. M. Gelfand, E. G. Glagoleva, E. B. Dynkin, V. S. Ryabenkii, B. V. Shabat, and also to the students of grades 11-D and 10-D of School No. 7 and the students of grades 10-E and 10-H of School No. 2 (Moscow), particularly M. Podolnii, A. Ryskin, E. Shats, and M. Shifrin.

# Contents


**Foreword**

| | *Problems* | *Solutions* | *Answers and Hints* |
|---|---|---|---|
| CHAPTER 1 | | | |
| **Sequences** | **1** | **42** | **132** |
| 1. What Are Sequences? | *1* | | |
| 2. The Method of Mathematical Induction | *4* | | |
| 3. Problems | *8* | | |
| CHAPTER 2 | | | |
| **Combinations** | **14** | **74** | **134** |
| CHAPTER 3 | | | |
| **Limits** | **20** | **96** | **137** |
| 1. Introductory Problems | *20* | | |
| 2. Problems Related to the Definition of Limit | *25* | | |
| 3. Problems on the Computation of Limits | *30* | | |
| **Test Problems** | | | |
| For Chapter 1 | *36* | | |
| For Chapter 2 | *38* | | |
| For Chapter 3 | *40* | | |

# SEQUENCES, COMBINATIONS, LIMITS

CHAPTER 1

# Sequences

## 1. What Are Sequences?

We shall say that we have a *numerical sequence*

$$u_1, u_2, \ldots, u_n, \ldots$$

if to each natural number[1] $n$ there corresponds a number $u_n$.

Here are some examples of sequences:

**Example 1.** $1, 1, 1, \ldots, 1, \ldots.$

Here $u_n = 1$; that is, the number 1 corresponds to each $n$.

**Example 2.** A sequence of odd integers:

$$1, 3, 5, \ldots.$$

Here $u_n = 2n - 1$.

**Example 3.** A sequence is given by the formula

$$u_n = \frac{n(n+1)}{2}.$$

(a) Write out this sequence.

(**Answer:** 1, 3, 6, 10, 15, 21, 28, . . . .)

[1]Positive integers are called *natural numbers*.

(b) Find $u_{100}$, $u_{n-3}$, and $u_{n+1}$.

(**Answer:** $u_{100} = 5050$;

$$u_{n-3} = \frac{(n-3)(n-3+1)}{2} = \frac{(n-3)(n-2)}{2};$$

$$u_{n+1} = \frac{(n+1)(n+2)}{2}.)$$

The formula giving $u_n$ is called the *formula of the general term* of the sequence. For example, the sequence

$$1, 4, 9, 16, 25, 36, 49, \ldots$$

can be written in abbreviated form using the formula for the general term: $u_n = n^2$.

**Example 4.** Write down the formula of the general term of the following sequence:

$$2, 5, 8, 11, 14, 17, \ldots.$$

**Example 5.** The law according to which the terms of a sequence are written down is not always easy to discover. For example, suppose that we have the following sequence:

$$0, \tfrac{7}{2}, 13, \tfrac{63}{2}, 62, \tfrac{215}{2}, 171, \ldots.$$

Find a rule that defines this sequence.

(**Answer.** A formula for the general term of this sequence has, for example, the form:

$$u_n = \frac{n^3 - 1}{2}.)$$

**Example 6.** For the sequence

$$3, 1, 4, 1, 5, 9, 2, 6, 5, 3, 5, \ldots$$

it is scarcely possible to write down a reasonable formula for the general term. There is, nevertheless, a rule defining the sequence: The $n$th term is the $n$th digit of the number $\pi$ written in the form of a decimal fraction ($\pi = 3.1415926\ldots$[1]). Thus we have a definite rule assigning to every natural number $n$ a number $u_n$ the $n$th term of the sequence. For example, $u_1 = 3$, $u_2 = 1$, $u_7 = 2$.

In principle we can find not only the first or the seventh term, but also the 100th term, the 105th term, or any other term. Although this sequence is not defined by a formula for the general term, one can establish a number of interesting facts about it.

Mathematicians have proved, for example, that it is not periodic.

**Example 7.** Write down the first ten terms of the sequence

$$1, 1, 2, 3, 5, \ldots,$$

which is given by the following rule: The first two of its terms are equal to one ($u_1 = 1$, $u_2 = 1$), and each term beginning with the third is equal to the sum of the preceding two. Clearly, the sequence is given by the formula

$$u_n = u_{n-1} + u_{n-2} \qquad (n \geq 3).$$

In this case it is possible to write down an explicit expression for $u_n$, but this is not very easy to do. (The terms of this sequence are called the *Fibonacci numbers*. See Problem 43.)

[1]You know from geometry that the number $\pi$ is the ratio of the length of the perimeter of a circle to its diameter. The first value of $\pi$ was found by Archimedes, who knew that $\pi$ is between $3\frac{1}{7}$ and $3\frac{10}{71}$. For solving problems in school you usually use the value $\pi = 3.14$. For more refined computations one takes a more exact value of $\pi$, for example, 3.14159. One can in principle compute the value of $\pi$ to any degree of accuracy. But it is difficult to conceive of a problem where one would need to know more than the first 10 digits or so. It is a curious fact that a man by the name of Shanks spent a great part of his life computing $\pi$ to 707 digits. Not long ago an electronic computer was used to find $\pi$ to several tens of thousands of places and it was found that Shanks's results were correct only to the first 202 places.

**Example 8.** Find the first terms of the sequence

$$u_1, u_2, \ldots,$$

the $n$th term of which is equal to the sum of all the natural numbers from 1 to $n$ inclusive.

(**Solution:** $u_1 = 1$, $u_2 = 1 + 2 = 3$, $u_3 = 1 + 2 + 3 = 6$, $u_4 = 1 + 2 + 3 + 4 = 10, \ldots,$ $u_n = 1 + 2 + 3 + \ldots + n$.

**Answer.** 1, 3, 6, 10, 15, 21, . . . .)

## 2. The Method of Mathematical Induction

If the first person in a line is a woman and the person behind each woman in the line is a woman, then everyone in the line is a woman.

The argument contained in this comic example is frequently met with in the most varied branches of mathematics and is called the *principle of mathematical induction.* A more serious formulation of this principle is the following:

*Suppose one has a sequence of assertions. If the first assertion is true, and if each true assertion is followed by a true assertion, then all the assertions in the sequence are true.*

**Example 9.** Suppose one wants to prove the formula

$$1 + 2 + 3 + \ldots + n = \frac{n(n+1)}{2}.$$

This formula is actually a whole sequence of assertions:

1. $$1 = \frac{1 \cdot 2}{2},$$

2. $$1 + 2 = \frac{2 \cdot 3}{2},$$

3. $$1 + 2 + 3 = \frac{3 \cdot 4}{2},$$

$$4.\ 1 + 2 + 3 + 4 = \frac{4 \cdot 5}{2},$$

The first assertion is, of course, true. Let us prove that the assertion following each true assertion is also true.

Suppose the $k$th assertion is true; that is,

$$1 + 2 + 3 + \ldots + k = \frac{k(k+1)}{2}.$$

Let us add the number $k + 1$ to both sides of the equation. We obtain

$$1 + 2 + 3 + \ldots + k + (k + 1) = \\ = \frac{k(k+1)}{2} + k + 1 = \frac{(k+1)(k+2)}{2}.$$

But this is the $(k + 1)$st assertion, which is thus true if the $k$th one is.

We have thus shown that the assertion following each true assertion is also true. By virtue of the principle of mathematical induction, our formula is true for all $k$.

This particular problem can be solved even without the use of the principle of mathematical induction. Let us denote the sum that we want to find by $u_n$. Let us write down two equations

$$u_n = 1 + 2 + 3 + \ldots + (n - 2) + (n - 1) + n,$$
$$u_n = n + (n - 1) + (n - 2) + \ldots + 3 + 2 + 1.$$

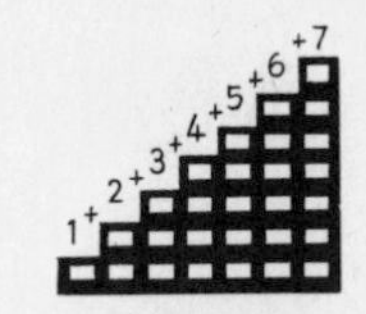


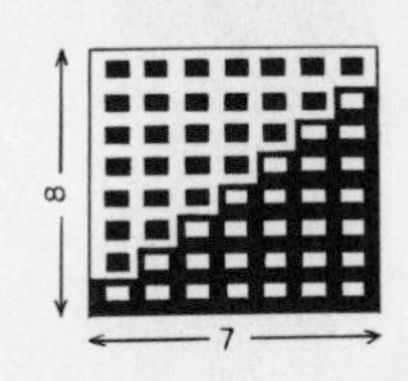


$1+2+3+4+5+6+7 = \frac{8 \cdot 7}{2} = 28$

Fig. 1

(The sum in the second line is the same as that in the first but written in the reverse order.) We get

$$2u_n = [1 + n] + [2 + (n - 1)] + \\ + [3 + (n - 2)] + \ldots + [(n - 2) + 3] \\ + [(n - 1) + 2] + [n + 1].$$

Each of the brackets contains the number $n + 1$; and there are $n$ such brackets in all (Fig. 1).

Therefore,

$$2u_n = \underbrace{(n+1) + (n+1) + \ldots + (n+1) + (n+1)}_{n \text{ times}} =$$

$$= n(n+1),$$

$$u_n = \frac{n(n+1)}{2},$$

and our equation is proved.

Let us now introduce another formulation of the principle of mathematical induction, somewhat different from the first.

Suppose that we have some proposition (also called a hypothesis) and that we want to prove the validity of this hypothesis for all natural numbers $n$.

Suppose that we have succeeded in proving the following two propositions:

(a) Our hypothesis is valid for $n = 1$.

(b) From the assumption that the hypothesis is valid for $n = k$, where $k$ is an arbitrary natural number, it follows that it is also true for $n = k + 1$.

The *principle of mathematical induction* is the assertion that the validity of statements (a) and (b) implies the validity of our hypothesis for all natural numbers $n$.

Statement (b) is conditional. It does not assert that the hypothesis is true for $n = k + 1$. It asserts only that *if* our hypothesis is true for $n = k$, then it is also true for $n = k + 1$. We shall call (b) the *inductive assumption.*

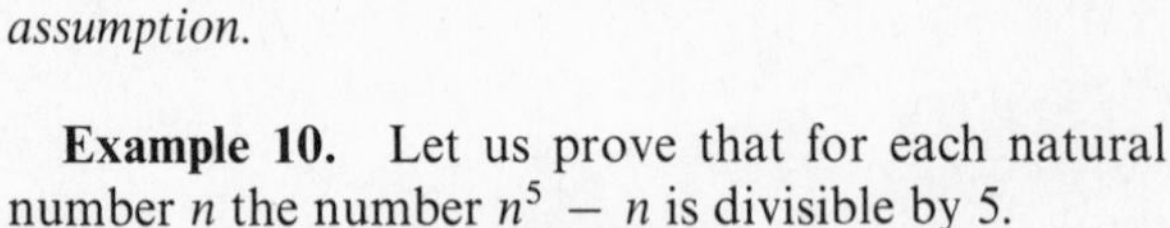

**Example 10.** Let us prove that for each natural number $n$ the number $n^5 - n$ is divisible by 5.

We carry out the proof using the method of mathematical induction.

(a) For $n = 1$ the expression $n^5 - n$ is equal to zero and therefore is divisible by 5.

(b) Let $k$ be an arbitrary natural number. We shall show that if for $n = k$ the number $n^5 - n$ is divisible by 5, then $(k + 1)^5 - (k + 1)$ is also divisible by 5.

Using the equality

$$(k + 1)^5 = k^5 + 5k^4 + 10k^3 + 10k^2 + 5k + 1$$

(check it), we represent the number $(k + 1)^5 - (k + 1)$ in the form

$$\begin{aligned} &(k + 1)^5 - (k + 1) = \\ &= (k^5 + 5k^4 + 10k^3 + 10k^2 + 5k + 1) - (k + 1) \\ &= (k^5 - k) + 5(k^4 + 2k^3 + 2k^2 + k). \end{aligned}$$

We have thus represented the number $(k + 1)^5 - (k + 1)$ as the sum of two terms. The first of these, $k^5 - k$, is divisible by 5, by our assumption. The second term, $5(k^4 + 2k^3 + 2k^2 + k)$, is likewise divisible by 5. The sum of two numbers divisible by 5 is divisible by 5; and so $(k + 1)^5 - (k + 1)$ is divisible by 5.

We have thus proved assertions (a) and (b). Therefore, by the principle of mathematical induction, the number $n^5 - n$ is divisible by 5 for all natural numbers $n$.

There exist other formulations of the principle of mathematical induction equivalent to those we have given. Here is one of them:

If (a) an assertion is true for $n = 1$; and (b) from the fact that it is true for all $n \leq k$, it follows that it is true for $n = k + 1$ as well, then the assertion is true for all $n$.

The difference between this formulation and the preceding one is the following.

In (b) we demand that our assertion be valid for all $n \leq k$ and not only for $n = k$.

It is easy to see that both formulations of the principle of mathematical induction are equivalent in the sense that every theorem that can be proved by means of the method of induction in one of its forms can be proved using the other form.

Let us note further that the proof of assertion (a) is needed, so to speak, merely for "priming." If instead we prove assertion

(a′) that our hypothesis is true for, say, $n = 8$, then (a′) and (b) will imply that our hypothesis is true for all natural numbers beginning with 8. We shall meet an example of this situation in Problem 9.

## 3. Problems

**1.** Prove that for all natural numbers $n$ the following equality (Fig. 2) is satisfied:

$$1^3 + 2^3 + 3^3 + \dots + n^3 = \frac{n^2(n+1)^2}{4}.$$

Fig. 2

**2.** Prove that for all natural numbers $n$ the following equality is satisfied:

$$\frac{1}{1\cdot 2} + \frac{1}{2\cdot 3} + \dots + \frac{1}{n(n+1)} = \frac{n}{n+1}.$$

**3.** Compute the sum

$$\frac{1}{1\cdot 4} + \frac{1}{4\cdot 7} + \dots + \frac{1}{(3n-2)(3n+1)}.$$

**4.** Prove that the equality

$$\frac{1}{a(a+1)} + \frac{1}{(a+1)(a+2)} + \dots + \frac{1}{(a+n-1)(a+n)} = \frac{n}{a(a+n)}$$

is valid for all natural numbers $n$ and all $a$ which are not equal to zero or to a whole negative number.

**5.** Prove the equality

$$1\cdot 1! + 2\cdot 2! + 3\cdot 3! + \dots + n\cdot n! = (n+1)! - 1$$[1]

(where $n$ is an arbitrary natural number).

**6.** Prove that the equality

$$\left(1 - \frac{1}{4}\right)\left(1 - \frac{1}{9}\right)\left(1 - \frac{1}{16}\right)\dots\left(1 - \frac{1}{n^2}\right) = \frac{n+1}{2n}$$

is valid for all whole numbers $n \geq 2$.

[1]By $n!$ (read "$n$ factorial") we mean the product of all integers from 1 to $n$.

**7.** Prove the identity:

$$1 - 2^2 + 3^2 - 4^2 + \ldots + (-1)^{n-1} n^2 = (-1)^{n-1} \frac{n(n+1)}{2}.$$

**8.** Into how many parts do $n$ lines divide the plane if no two of the lines are parallel and no three of them intersect in a point?

**9.** Prove that any sum of money in excess of 7 kopeks can be paid in three- and five-kopek pieces (Fig. 3).

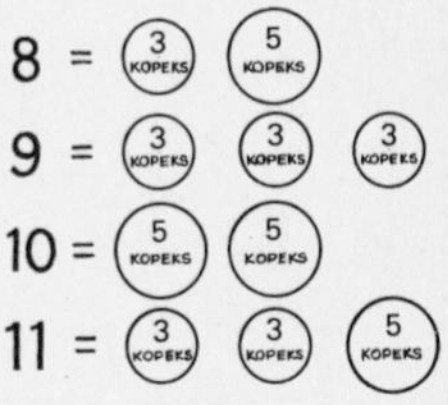

Fig. 3

**10.** Prove that a plane divided into $n$ parts by straight lines can be painted using black and white paint in such a way that any two parts having a common boundary will be painted different colors (such a design is called regular; Fig. 4).

**11.** Prove that the sum of the cubes of any three consecutive natural numbers is divisible by 9.

**12.** Prove that for each whole number $n \geq 0$

$$11^{n+2} + 12^{2n+1}$$

is divisible by 133.

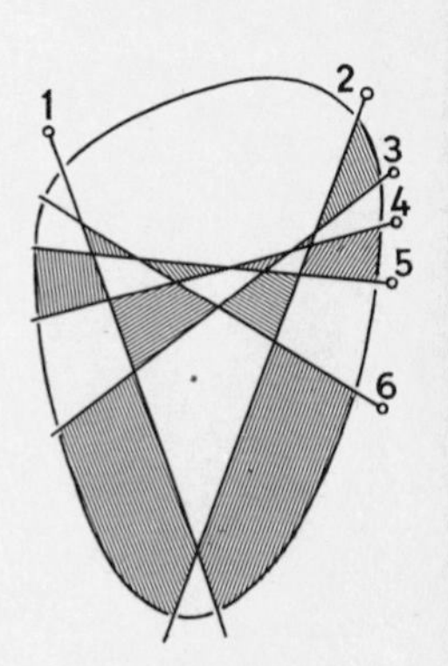

Fig. 4

**13.** Prove that for any $n \geq 2$ the inequality

$$(1 + a)^n > 1 + na$$

is valid if $a > -1$ and $a \neq 0$.

**14.** Prove that for each $n \geq 2$

$$1 + \frac{1}{\sqrt{2}} + \frac{1}{\sqrt{3}} + \ldots + \frac{1}{\sqrt{n}} > \sqrt{n}.$$

We shall now turn to problems on sequences. It will be useful to re-examine the definition of sequences and the examples in Section 1.

**15.** Let a sequence be given in the following manner:

$$u_1 = 2; \quad u_2 = 3; \quad u_n = 3u_{n-1} - 2u_{n-2} \text{ for } n > 2.$$

Prove that the formula $u_n = 2^{n-1} + 1$ is valid.

**16.** Let $u_1 = 1; u_{n+1} = u_n + 8n$, where $n$ is a natural number. Prove that $u_n = (2n-1)^2$.

Assume that a sequence $u_1, u_2, \ldots$ is given. Define a new sequence $v_1, v_2, \ldots$ in the following manner: $v_n = u_{n+1} - u_n$ for all $n$. This sequence is called the sequence of differences of the sequence $u_1, u_2, \ldots$. We denote our new sequence as follows: $\Delta u_1, \Delta u_2, \ldots$ ($\Delta$ is the capital Greek letter *delta*).

**17.** Let $u_n = n^2$. Find the formula for the general term of the sequence of differences.

**18.** Suppose that we have two sequences $u_1, u_2, \ldots$ and $v_1, v_2, \ldots$ with the same sequences of differences; that is, $\Delta u_n = \Delta v_n$ for all $n$.

(a) Can we assert that $u_n = v_n$?

(b) Can we assert that $u_n = v_n$ if we also know that $u_1 = v_1$?

**19.** Assume that each term of the sequence $w_1, w_2, \ldots$ is equal to the sum of the corresponding terms of the sequences $u_1, u_2, \ldots$ and $v_1, v_2, \ldots$; that is, $w_n = u_n + v_n$ for all $n$. Prove that in this case $\Delta w_n = \Delta u_n + \Delta v_n$ for all $n$.

**20.** Given the sequence $u_n = n^k$, show that the general term of its sequence of differences is a polynomial in $n$ of degree $k - 1$ and find the leading coefficient of this polynomial.

**21.** Given a sequence for which $u_n$ is a polynomial in $n$ of degree $k$ with leading coefficient $a_0$, prove that $\Delta u_n$ is a polynomial of degree $k - 1$ in $n$ and find its leading coefficient.

**22.** Given the sequence $u_n$ $n^k$ ($k$ is a fixed positive number), find the sequence of differences, determine the sequence of differences of this new sequence, and do this $k$ times (Fig. 5). Prove that one obtains a sequence all of whose terms are equal to the same number. What is this number?

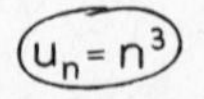


| $n$ | $n^3$ | 1ST difference | 2ND difference | 3RD difference |
|---|---|---|---|---|
| 0 | 0 | | | |
| | | 1 | | |
| 1 | 1 | | 6 | |
| | | 7 | | 6 |
| 2 | 8 | | 12 | |
| | | 19 | | 6 |
| 3 | 27 | | 18 | |
| | | 37 | | 6 |
| 4 | 64 | | 24 | |
| | | 61 | | 6 |
| 5 | 125 | | 30 | |
| | | 91 | | |
| 6 | 216 | | | |
| ⋮ | ⋮ | ⋮ | ⋮ | ⋮ |

Fig. 5

**23.** Given the sequence $v_n = n^k$, prove that there exists a sequence $u_1, u_2, \ldots$ whose general term is a polynomial of degree $k + 1$ in $n$ and for which $\Delta u_n = v_n$. Also, find the leading coefficient of this polynomial.

**24.** Suppose that for the sequence $u_1, u_2, \ldots$ the sequence of differences has been found and that the sequence of differences of this new sequence has likewise been found. Suppose, in addition, that this process has been carried out four times and that the resulting sequence consists entirely of zeros. Prove that the general term of the initial sequence is given by a polynomial of the third degree.

**25.** It is known that the sequence of differences $v_1, v_2, \ldots$ of the sequence $u_1, u_2, \ldots$ is given by a polynomial of degree $k$. Prove that the sequence $u_1, u_2, \ldots$ can be given by a polynomial of degree $k + 1$.

**26.** Compute the sum $1^2 + 2^2 + \ldots + n^2$.

**27.** Find the sum

$$1 \cdot 2 + 2 \cdot 3 + 3 \cdot 4 + 4 \cdot 5 + \ldots + n(n + 1).$$

An *arithmetic progression* is a sequence for which $u_{n+1} = u_n + d$ for all $n$. The number $d$ is called the *increment* of the progression.

A *geometric progression* is a sequence for which $u_{n+1} = u_n \cdot q$ for all $n$. The number $q$ is called the *ratio* of the progression.

**28.** (a) Express the $n$th term of an arithmetic progression in terms of the first term $u_1$ and the increment $d$.

(b) Express the $n$th term of a geometric progression in terms of the first term $u_1$ and the ratio $q$.

**29.** (a) Find a formula for the sum $S_n$ of the first $n$ terms of an arithmetic progression.

(b) Find a formula for the product $P_n$ of the first $n$ terms of a geometric progression.

**30.** (a) Find the sum of the first 15 terms of an arithmetic progression for which the first term is 0 and the increment is $\frac{1}{3}$.

(b) Find the product of the first 15 terms of the geometrical progression for which the first term is 1 and the ratio is $\sqrt[3]{10}$.

**31.** (a) The third term of an arithmetic progression is equal to 0. Find the sum of the first 5 terms.

(b) The third term of a geometric progression is

equal to 4. Find the product of the first 5 terms.

**32.** Prove that the sum $S_n$ of the first $n$ terms of a geometric progression is given by the formula $S_n = a_1(q^n - 1)/(q - 1)$.

**33.** Someone arrives in a city with very interesting news and within 10 minutes tells it to two others. Each of these tells the news within 10 minutes to two others (who have not heard it yet), and so on.[1] How long will it take before everyone in the city has heard the news if the city has three million inhabitants?

**34.** A cyclist and a horseman have a race in a stadium. The course is five laps long. They spend the same time on the first lap. The cyclist travels each succeeding lap 1.1 times more slowly than he does the preceding one. On each lap the horseman spends $d$ minutes more than he spent on the preceding lap. They each arrive at the finish line at the same time. Which of them spends the greater amount of time on the fifth lap and how much greater is this amount of time?

**35.** Find the sum of all positive odd numbers not exceeding one thousand.

**36.** Find the sum of all positive three-digit numbers not divisible by either 2 or 3.

**37.** The sum $S_n$ of the first $n$ terms of a sequence is expressed by the formula $S_n = 3n^2$. Prove that this sequence is an arithmetic progression. Compute its first term and its increment.

**38.** Does there exist a geometric progression among whose terms the numbers 27, 8, and 12 appear? (These numbers need not be consecutive and need not appear in the given order.) In what positions can these numbers appear in a progression?

**39.** The same question for the numbers 1, 2, 5.

**40.** The squares of the 12th, 13th, and 15th terms of an arithmetic progression form a geometric progression. Find all of the possible ratios of this progression.

**41.** Find all the geometric progressions for which

[1] As long, of course, as there are people left in the city who do not yet know the news.

each term beginning with the third is equal to the sum of the preceding two.

**42.** The terms of a sequence are the sums of the corresponding terms of two geometric progressions. What is the third term of this sequence if the first two terms are zero?

**43.** Represent the terms of the Fibonacci sequence ($u_1 = 1; u_2 = 1; u_{n+2} = u_{n+1} + u_n$)[1] in the form of the sum of the corresponding terms of two geometric progressions.

[1]See Example 7 on page 3.

CHAPTER 2

# Combinations

Most of the problems in this chapter deal with the question "How many ways?" Problems of this kind (on the computation of the number of possible combinations) are frequently called combinatorial; and the branch of mathematics dealing with their solutions is called *combinatorial analysis.* Combinatorial calculations have great importance in the theory of probability, numerical mathematics, the theory of automata, and mathematical economics.

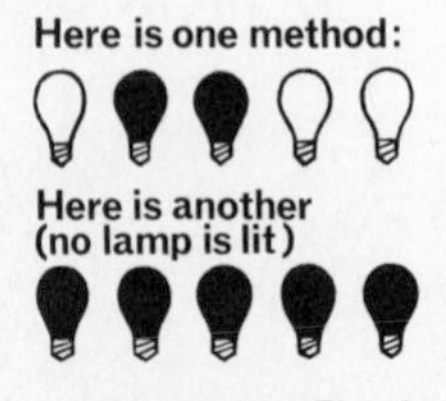


Fig. 6

**44.** In a kitchen there are five lamps. In how many ways can the kitchen be lighted?

Is it clear what methods of lighting there are? Each lamp is either lit or not lit. Two ways are considered different if they differ in the state of a single lamp (Figs. 6 and 7).

**45.** Suppose that we have $n$ lamps. We denote by $C_n^k$ the number of ways of lighting such that $k$ lamps are lit. Prove that $C_n^0 + C_n^1 + \ldots + C_n^k + \ldots + C_n^n = 2^n$. Is it clear what the symbol $C_n^k$ means? What does $C_5^1$ mean? What is $C_5^4$ equal to?

**46.** There are $n$ traffic lights in a city. Each can be in one of three states (red, yellow, or green). In how many ways can all of the traffic lights be lit?

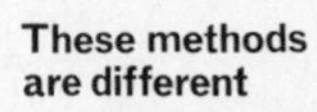

These are
also different

Fig. 7

**47.** In how many ways can $n$ traffic lights be lit if $k$ of them can be in one of three states and the remaining $n - k$ in one of two?

**48.** How many six-digit numbers are there not containing 0 or 8?

**49.** What is the maximum possible number of different automobile license numbers consisting of four digits followed by three letters using an alphabet of 26 letters)?

**50.** In a certain kingdom each person has a unique set of teeth. What is the maximum number of inhabitants of the kingdom (the maximum number of teeth is 32)?

**51.** In the expression $(x - 1)(x - 2)(x - 3) \dots (x - 100)$ the parentheses are removed and the multiplication carried out. Find the coefficient of $x^{99}$.

**52.** In how many ways can the natural number $n$ be represented in the form of a sum of two natural numbers?

This problem can be interpreted in alternate ways, depending upon whether or not you consider the order of the summands essential. In other words, you can regard the representations $8 = 3 + 5$ and $8 = 5 + 3$ either as identical or as distinct. The answer obtained will, of course, depend upon your interpretation. Solve both problems.

**53.** The analogous question with three summands.

*Pascal's triangle* is a table of the following form:

$$\begin{array}{ccccccccccc} &&&&& 1 &&&&& \\ &&&& 1 && 1 &&&& \\ &&& 1 && 2 && 1 &&& \\ && 1 && 3 && 3 && 1 && \\ & 1 && 4 && 6 && 4 && 1 & \\ 1 && 5 && 10 && 10 && 5 && 1 \end{array}$$

(the sides of the triangle consist of ones, and each of the remaining numbers in the table is the sum of the two numbers above it).

**54.** Prove that in Pascal's triangle the sum of the numbers in the $(n + 1)$st line is equal to $2^n$.

**55.** What is the greatest number of bishops that can be placed on a chessboard in such a way that they do not threaten one another? Prove that the number of different ways that this can be done is a square.

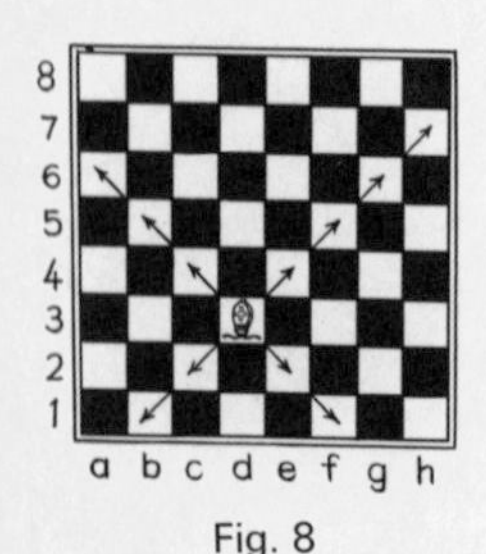


Fig. 8

To understand the conditions, one must know, of course, that a bishop moves only along diagonals. For example, a bishop located on square $d3$ can in one move reach any of the squares marked off in Fig. 9 ($b1$, $c2$, and so on). Thus, two bishops threaten one another only if they are located on the same diagonal.

**56.** A woman has two apples and three pears. For five consecutive days she gives away one piece of fruit. In how many ways can she do this?

**57.** The analogous problem for $k$ apples and $n$ pears.

**58.** The analogous problem for 2 apples, 3 pears, and 4 oranges.

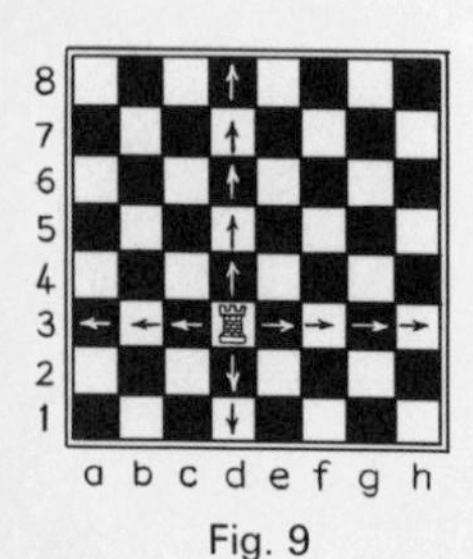


Fig. 9

**59.** In how many ways can one set up 8 rooks on a chessboard so that they do not threaten one another? Rooks move only along horizontals and verticals. For example, the rook on $d3$ threatens each square of the column $d$ and the row 3 (Fig. 9).

**60.** In how many ways can one choose two objects from $n$?

We have denoted the number of ways of choosing two objects from $n$ (or of lighting a kitchen if two lamps out of $n$ are to be lit) by $C_n^2$. The problem thus consists in computing the number $C_n^2$ (for example, in Problem 56 we found that $C_5^2 = 10$).

**61.** In how many ways can one seat a class if there are 26 people and 28 seats?

**62.** A father has five distinct oranges which he gives to his eight sons so that each one receives either one orange or none. In how many ways can this be done?

**63.** We denote by $C_n^k$ the number of ways in which one can have $k$ out of $n$ lamps lit (each lamp can be in either of the two states). Prove that the number $C_n^k$ is the number in the $(k + 1)$st place of the $(n + 1)$st line of Pascal's triangle.

**64.** In how many ways can one place 4 rooks on a chessboard so that they do not threaten one another?

**65.** In a company there are 3 officers and 40 soldiers. In how many ways can a detail consisting of one officer and three soldiers be chosen?

**66.** In how many ways can one choose three objects out of $n$?

**67.** In how many ways can one choose $k$ objects out of $n$?

**68.** A piano has 88 keys. How many sequences of 6 distinct sounds are there? (In a sequence the sounds go one after the other.) How many chords of six sounds are there? (A chord is obtained by pressing six keys simultaneously.)

**69.** In which rows of Pascal's triangle are all of the numbers odd?

**70.** How many terms are obtained after multiplying out the expression

$$(a + 1)(b + 1)(c + 1)(d + 1)(e + 1)(f + 1)(g + 1)?$$

**71.** How many of the terms will consist of three letters?

**72.** The expression $(1 + x + y)^{20}$ is expanded but like terms are not reduced. How many terms are obtained?

**73.** The expression $(1 + x^5 + x^7)^{20}$ is expanded. Determine the coefficients of $x^{17}$ and $x^{18}$ after the reduction of like terms.

**74.** The expression $(1 + x)^{56}$ is expanded and like terms reduced. Find the coefficients of $x^8$ and $x^{48}$.

**75.** Prove that

$$(1 + x)^n = C_n^0 + C_n^1 x + C_n^2 x^2 + \dots + C_n^{n-1} x^{n-1} + C_n^n x^n.$$

This is *Newton's binomial formula.*

**76.** From Problems 75 and 63 we can deduce the following result. Let $a_0, a_1, a_2, \dots, a_n$ be the numbers in the $(n + 1)$st line of Pascal's triangle. Then

$$(1 + x)^n = a_0 + a_1 x + a_2 x^2 + \dots + a_n x^n.$$

Prove this assertion, relying directly upon the definition of Pascal's triangle (and not using the results of Problems 75 and 63).

**77.** Between each pair of digits of the number 14641, $k$ zeros are inserted. Prove that the number obtained is a perfect square.

**78.** The expression $(a + b)^n$ is expanded and like terms are reduced. Write down the term containing $a^k$.

**79.** In the expression $(x + y + z)^n$ find the term containing $x^k y^l$.

**80.** Find (a) $C_n^0 + C_n^2 + C_n^4 + C_n^6 + \dots$;

(b) $C_n^1 + C_n^3 + C_n^5 + \dots$.

**81.** Determine the sum of the coefficients of the polynomial that is obtained if we expand the expression $(1 + x - 3x^2)^{1965}$ and reduce like terms.

**82.** How many zeros does the number $11^{100} - 1$ end in?

**83.** Which is greater: $99^{50} + 100^{50}$ or $101^{50}$?

**84.** How many distinct divisors does the number $2 \cdot 3 \cdot 5 \cdot 7 \cdot 1$ have?

**85.** How many distinct divisors does the number $10!$ have?

**86.** How many diagonals does a convex $n$-gon have?

**87.** How many different even five-digit numbers are there consisting of the digits 0, 1, 3, 4, 5, with no digits repeated?

**88.** In how many ways can one put eight coins of different values in two pockets?

**89.** Ten subjects are studied in class. On Wednesday there are six lessons, each of which is different. In how many ways can one establish a schedule for Wednesday?

**90.** In how many ways can one distribute $3n$ objects among three people so that each one receives $n$ of the objects?

**91.** Some people in a room know at least one of three languages. Six know English, six know German, and seven know French. Four know English and

German. Three know German and French. Two know French and English. One person knows all three languages. How many people are in the room? How many of them know only English?

**92.** A library has a certain number of readers (that is, people who have read at least one of the library's books. For any $k$ books ($1 \le k \le n$) we know how many people have read them all. How do we find the total number of readers? (The library has $n$ books in all.)

**93.** How many telephone numbers are there containing the combination 12? (A number consists of six digits.)

**94.** The same question for numbers of $n$ digits.

CHAPTER 3

# Limits

This chapter contains more supplementary material and demands much independent work of you.

The problems here can be divided into two groups. The first group consists of the problems in an introductory section (except for some difficult problems, marked with asterisks) and the problems of the main section, which are marked with circles. Those of you who desire to master the theory of limits in the framework of the high school curriculum can limit yourselves to the problems in the first group. Test Problems 33, 34a, 34b, 35, 39, 40a, 40d, and 43 will enable you to determine how well you have learned the material.

The problems in the second group, which are an introduction to mathematical analysis, presuppose greater preparation in logic. Independent work is particularly important in the solution of these problems. Here, we do not recommend that you use the passive method of study mentioned in the introduction.

## 1. Introductory Problems

**95.** Two students were given an assignment to prepare a calendar of the weather. Each day they are to mark down a "+" if the weather is good, and "−" if the weather is bad. The first student proceeds in the following manner. He makes observations three times a day—in the morning, the afternoon, and the evening. If it is raining during at least one observation, he puts a minus sign on the calendar. Otherwise, he puts down a plus sign. The second student makes his observations at the same times as the first. If it is not raining during at least one observation, he puts down a "+." In the other cases he puts down a "−." Thus each day the weather receives one of the following marks: ++; +−; −+; −−. Are all of these marks actually possible?

**96.** A third student joins the students in Problem 95; he makes observations at the same times as the first two and places a "−" if it is raining during at least two observations and a "+" otherwise. Which of the eight marks: +++; ++−; +−+; −++; −+−; −−+; +−−; −−− can in fact occur?

**97.** (a) Three hundred people stand in a rectangular array of 30 rows and 10 columns. In each row the tallest person is chosen; and from these 30 people the shortest is chosen. Then the shortest person in each column is chosen; and from these 10 the tallest is chosen. Who is taller: the tallest of the shortest or the shortest of the tallest?

(b) Is the answer changed if the people line up as in Fig. 10 (20 people in the first ten columns and 10 in the remainder) instead of in a rectangle?

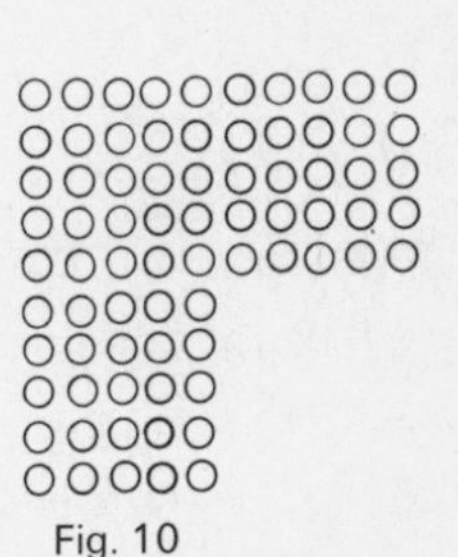
Fig. 10

**98.** A test is called easy if at each desk there is a student who has solved all of the problems. Formulate a definition of a difficult test.

**99.** Let us consider two definitions of an easy test:

(a) Each time the test is given, each problem is solved by at least one student.

(b) Each time the test is given, at least one student solves all of the problems.

Can a test be easy in the sense of Definition (a) and difficult in the sense of Definition (b)?

**100.** Which of the following theorems is true?

1. If each summand is divisible by 7, then the sum is divisible by 7.

2. If each summand is indivisible by 7, then the sum is not divisible by 7.

3. If at least one summand is divisible by 7, then the sum is divisible by 7.

4. If a sum is divisible by 7, then each summand is divisible by 7.

5. If a sum is not divisible by 7, then each summand is indivisible by 7.

6. If a sum is not divisible by 7, then at least one summand is not divisible by 7.

**101*.** Let **A** and **B** denote two propositions. A bar over a letter will denote the negation of the corresponding proposition (Fig. 11). For example, if the letter **A** denotes the proposition "In triangle *ABC* all of the sides are equal," then $\bar{\mathbf{A}}$ will denote the proposition "In triangle *ABC* not all of the sides are equal." Consider the following eight theorems:

1. If **A**, then **B**.
2. If $\bar{\mathbf{A}}$, then **B**.
3. If **A**, then $\bar{\mathbf{B}}$.
4. If $\bar{\mathbf{A}}$, then $\bar{\mathbf{B}}$.
5. If **B**, then **A**.
6. If $\bar{\mathbf{B}}$, then **A**.
7. If **B**, then $\bar{\mathbf{A}}$.
8. If $\bar{\mathbf{B}}$, then $\bar{\mathbf{A}}$.

**EVENTS**

**A** -it's raining

$\bar{\mathbf{A}}$ -it's not raining

**B** -umbrella's open

$\bar{\mathbf{B}}$ -umbrella's not open

**THEOREMS**

**If A, then B**

**If $\bar{\mathbf{A}}$, then B**

**If A, then $\bar{\mathbf{B}}$**

**If $\bar{\mathbf{A}}$, then $\bar{\mathbf{B}}$**

Fig. 11

Assuming that Theorem 1 is true, divide the remaining theorems into three groups: first, theorems that are necessarily true; second, theorems that are necessarily false; and third, theorems that may be either true or false. Let us agree not to substitute for **A** and **B** propositions that are always false or always true (such as "In triangle *ABC* all of the angles are right angles" or "In triangles *ABC* the three medians intersect in a single point").

**102.** Solve the equations:[1]

(a) $x + 2|x| = 3$,

(b) $x^2 + 3|x| - 4 = 0$,

(c) $|2x + 1| + |2x - 1| = 2$.

**103.** Prove the inequalities:

(a) $|x + y| \leq |x| + |y|$,

(b) $|x - y| \geq |x| - |y|$,

(c) $|x - y| \geq \big||x| - |y|\big|$.

For each of these cases, determine when the inequality becomes an equality.

**104.** Is it true that there exist natural numbers $n$ such that

(a) $\sqrt[n]{1000} < 1.001$?

(b)* $\sqrt[n]{n} < 1.001$?

(c) $\sqrt{n + 1} - \sqrt{n} < 0.1$?

(d) $\sqrt{n^2 + n} - n < 0.1$?

**105.** Is it true that there exists a number $C$ such that for each integer $k$ the inequality

$$\left| \frac{k^3 - 2k + 1}{k^4 - 3} < C \right|$$

is satisfied?

[1]The quantity $|x|$ (read "the modulus of $x$" or the "absolute value of $x$") is defined in the following way:

$$|x| = \left.\begin{array}{r} x, \text{ if } x > 0 \\ 0, \text{ if } x = 0 \\ -x, \text{ if } x < 0. \end{array}\right\}$$

**106*.** Is it true that for each number $C$ there exists an infinite set of integers $k$ for which the inequality

$$k \sin k > C$$

is satisfied?

Suppose that we are given an infinite sequence $\{x_n\}$.[1] Let us represent the terms of this sequence by points on a number axis (certain terms of the sequence can, of course, coincide; in the sequence $1, \frac{1}{2}, 1, \frac{1}{3}, 1, \frac{1}{4}, \ldots$, for example, all of the odd-numbered terms are the same).

Let us call the segment $[a, b]$ on the real axis a *trap* for the sequence $\{x_n\}$ if there are at most a finite number of terms of the sequence which lie outside of the segment $[a, b]$.

Let us call the segment $[a, b]$ on a number axis a *trough* for the sequence $\{x_n\}$ if the segment $[a, b]$ contains an infinite number of terms of the sequence.

**107.** (a) Prove that every trap is a trough.

(b) Find a sequence and a segment such that the segment is a trough for the sequence but not a trap.

**108.** Suppose that we are given the sequences

(a) $1, \frac{1}{2}, \frac{1}{3}, \ldots, \frac{1}{n}, \ldots$,

(b) $1, 2, \frac{1}{2}, 1\frac{1}{2}, \frac{1}{3}, \ldots, \frac{1}{n}, 1\frac{1}{n}, \ldots$,

(c) $1, \frac{1}{2}, 3, \frac{1}{4}, 5, \frac{1}{6}, \ldots, 2n - 1, \frac{1}{2n} \ldots$

and the segments

(A) $[-\frac{1}{2}, \frac{1}{2}]$, (B) $[-1, 1]$, (C) $[-2, 2]$.

Determine which of these segments are traps or troughs for the sequences (a), (b), and (c).

**109.** Does a sequence exist for which each of the segments $[0, 1]$ and $[2, 3]$ is

(a) a trough?

(b) a trap?

**110.** Each of the segments $[0, 1]$ and $[9, 10]$ is a trough (for a certain sequence). Does this sequence have

(a) a trap of length 1?

(b) a trap of length 9?

[1]The symbol $\{x_n\}$ is short for the sequence $x_1, x_2, \ldots, x_n, \ldots$.

**111.** (a) Does there exist a sequence not having a single trough?

(b)* Does there exist a sequence for which every segment is a trough?

## 2. Problems Related to the Definition of Limit

The number $a$ is called the *limit of the sequence* $\{x_n\}$, if for any positive number $\varepsilon$ (the Greek letter "epsilon") there exists a number $k$ such that for all terms of the sequence for which the number $n$ is greater than $k$ the inequality

$$|x_n - a| < \varepsilon$$

is satisfied (Fig. 12).

Fig. 12

The fact that the number $a$ is the limit of the sequence $\{x_n\}$ is written as follows:

$$\lim_{n \to \infty} x_n = a$$

(this is read: "The limit of $x_n$ as $n$ approaches infinity is equal to $a$"), or

$$x_n \to a \quad \text{as} \quad n \to \infty$$

(this is read: "$x_n$ approaches $a$ as $n$ approaches infinity").

**112°.** Suppose that we are given the following sequences:

(a) $x_n = \dfrac{1}{n}$,

(b) $x_n = \dfrac{1}{n^2 + 1}$,

(c) $x_n = (-\frac{1}{2})^n$ (Fig. 13),

(d) $x_n = \log_n 2$.

In each of these cases find a number $k$ such that for $n > k$ the following inequalities are satisfied:

(A) $|x_n| < 1$, (B) $|x_n| < 0{\cdot}001$, (C) $|x_n| < 0{\cdot}000001$.

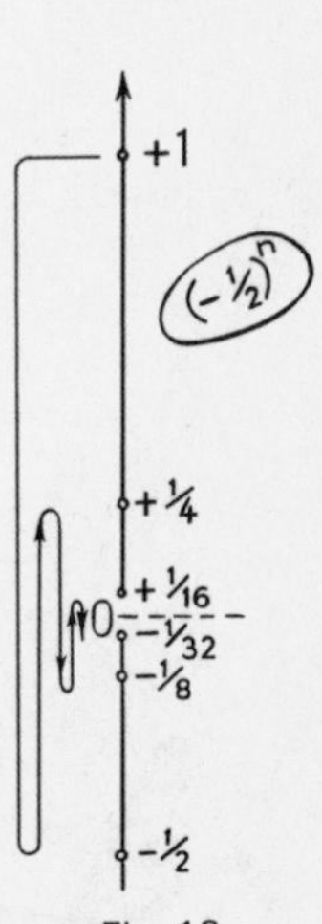

Fig. 13

**113.** (a) Prove that if $x_n \to a$ as $n \to \infty$, then each segment whose center is the point $a$ is a trap for the sequence $\{x_n\}$.

(b) Is the converse true?

**114*.** (a) Prove that if $x_n \to a$ as $n \to \infty$, then each segment whose center is the point $a$ is a trough (for the sequence $\{x_n\}$) and no segment that does not contain the point $a$ is a trough.

(b) It is known that for a certain sequence $\{x_n\}$ any segment with center at $a$ is a trough and that no segment which does not contain the point $a$ is a trough. Can one assert that $x_n \to a$ as $n \to \infty$?

**115.** Prove that if a certain segment is a trough for the sequence $\{x_n\}$, then no number outside this segment can be a limit of this sequence.

**116°.** Which of the following sequences have limits?

(a) $1, -\frac{1}{2}, \frac{1}{3}, -\frac{1}{4}, \ldots, \dfrac{1}{2n-1}, -\dfrac{1}{2n}, \ldots;$

(b) $0, \frac{2}{3}, \frac{8}{9}, \frac{26}{27}, \ldots, \dfrac{3^n - 1}{3^n}, \ldots;$

(c) $1, 1 + \frac{1}{2}, 1 + \frac{1}{2} + \frac{1}{4}, \ldots, 1 + \frac{1}{2} + \ldots + \dfrac{1}{2^n}, \ldots;$ (Fig. 16).

(d) $1, 2, 3, 4, \ldots, n, \ldots;$

(e) $1, 1, 1, 1, \ldots, 1, \ldots;$

(f) $0, 1, 0, \frac{1}{2}, 0, \frac{1}{3}, \ldots, 0, \frac{1}{n}, \ldots;$

(g) $0.2, 0.22, 0.222, \ldots /, \underbrace{0.222\ldots2}_{n}, \ldots;$

(h) $\sin 1°, \sin 2°, \ldots, \sin n°, \ldots;$

(i) $\dfrac{\cos 1°}{1}, \dfrac{\cos 2°}{2}, \dfrac{\cos 3°}{3}, \ldots, \dfrac{\cos n°}{n}, \ldots;$

(j) $0, 1\frac{1}{2}, -\frac{2}{3}, 1\frac{1}{4}, -\frac{4}{5}, 1\frac{1}{6}, \ldots, (-1)^n + \frac{1}{n}, \ldots;$ (Fig.

2

1 + ½ + ¼ + ⅛ + 1/16

1 + ½ + ¼ + ⅛

1 + ½ + ¼

1 + ½

1

Fig. 14

**117°.** Can two different numbers be limits of the same sequence?

**118.** The number $a$ is called a *limit point* of the sequence $\{x_n\}$ if for each positive number $\varepsilon$ and any number $k$ there exists a number

$$n > k,$$

such that the following inequality is satisfied:

$$|x_n - a| < \varepsilon.$$

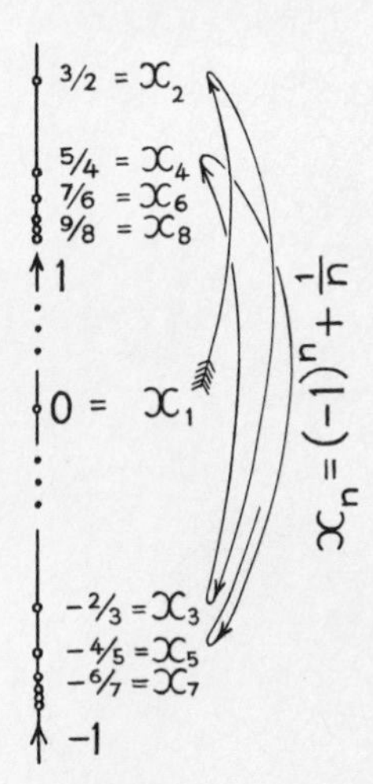


Fig. 15

(a) Prove that if $a$ is a limit point of the sequence $\{x_n\}$, then any segment having center $a$ is a trough for this sequence.

(b) Prove the converse.

**119.** Prove that the limit of a sequence (if it exists) is a limit point.

**120.** Find all of the limit points of each of the following sequences:

(a) $x_n = \dfrac{n+1}{n}$,

(b) $x_n = (-1)^n$,

(c) $x_n = \sin n^0$,

(d) $x_n = n^{(-1)n}$,

(e) $x_n = n$,

(f) $\frac{1}{2}, \frac{1}{3}, \frac{2}{3}, \frac{1}{4}, \frac{2}{4}, \frac{3}{4}, \frac{1}{5}, \frac{2}{5}, \frac{3}{5}, \frac{4}{5}, \ldots$.

**121.** The sequence $\{x_n\}$ is said to be bounded if there exists a number $C$ such that for any $n$ the following inequality is satisfied:

$$|x_n| \leq C.$$

Formulate the definition of an unbounded sequence.

**122.** (a) Prove that if a sequence has a limit, then it is bounded.

(b) Is the converse true?

**123.** One says that the sequence $\{x_n\}$ converges to infinity (this is written as follows: $x_n \to \infty$ as $n \to \infty$) if for any number $C$ there exists a number $k$ such that for all numbers $n > k$ the following inequality (Fig. 16) is satisfied:

$$|x_n| > C.$$

Fig. 16

Which of the following sequences converge to infinity and which of them are unbounded:

(a) $x_n = n$,

(b) $x_n = n \cdot (-1)^n$,

(c) $x_n = n^{(-1)^n}$,

(d) $x_n = \begin{cases} n \text{ for } n \text{ even}, \\ \sqrt{n} \text{ for } n \text{ odd}, \end{cases}$

(e) $x_n = \dfrac{100n}{100 + n^2}$?

**124.** Find a bounded sequence that
(a) has both a greatest and a least term,
(b) has a greatest term but not a least term,
(c) has a least term but not a greatest one,
(d) has neither a least term nor a greatest one.

**125.** Consider the following 16 conditions ("$\forall$" means "for all," "$\exists$" means "there exists," and "$\ni$" means "such that"):

1. $\exists\, \varepsilon > 0 \ni \exists\, k \ni \exists\, n > k \ni |x_n - a| < \varepsilon.$
2. $\exists\, \varepsilon > 0 \ni \exists\, k \ni \exists\, n > k \ni |x_n - a| > \varepsilon.$
3. $\exists\, \varepsilon > 0 \ni \exists\, k \ni \forall\, n > k, |x_n - a| < \varepsilon.$
4. $\exists\, \varepsilon > 0 \ni \exists\, k\, \forall\, n > k, |x_n - a| > \varepsilon.$
5. $\exists\, \varepsilon > 0 \ni \forall\, k\, \exists\, n > k \ni |x_n - a| < \varepsilon.$
6. $\exists\, \varepsilon > 0 \ni \forall\, k\, \exists\, n > k \ni |x_n - a| > \varepsilon.$
7. $\exists\, \varepsilon > 0 \ni \forall\, k$ and $\forall\, n > k, |x_n - a| < \varepsilon.$
8. $\exists\, \varepsilon > 0 \ni \forall\, k$ and $\forall\, n > k, |x_n - a| \geq \varepsilon.$
9. $\forall\, \varepsilon > 0, \exists\, k \ni \exists\, n > k \ni |x_n - a| < \varepsilon.$
10. $\forall\, \varepsilon > 0, \exists\, k \ni \exists\, n > k \ni |x_n - a| \geq \varepsilon.$
11. $\forall\, \varepsilon > 0, \exists\, k \ni \forall\, n > k, |x_n - a| < \varepsilon.$
12. $\forall\, \varepsilon > 0, \exists\, k \ni \forall\, n > k, |x_n - a| \geq \varepsilon.$
13. $\forall\, \varepsilon > 0$ and $\forall\, k, \exists\, n > k \ni |x_n - a| < \varepsilon.$
14. $\forall\, \varepsilon > 0$ and $\forall\, k, \exists\, n > k \ni |x_n - a| \geq \varepsilon.$
15. $\forall\, \varepsilon > 0, \forall\, k$, and $\forall\, n > k, |x_n - a| < \varepsilon.$
16. $\forall\, \varepsilon > 0, \forall\, k$, and $\forall\, n > k, |x_n - a| \geq \varepsilon.$

Which of these conditions expresses properties of sequences with which you are already familiar (boundedness, having a limit $a$, having a limit point $a$, converging to infinity) or the negation of these properties?

**126.** Consider the following five properties of sequences: (1) being identically equal to $a$, (2) having limit $a$, (3) having the number $a$ as limit point, (4) being bounded, (5) converging to infinity.

We can assign to each sequence a series of five plus or minus signs. The choice $- + + + -$, for example, will indicate that the sequence possesses properties 2, 3, and 4, but not 1 or 5. Certain choices, however, will not make sense (for example, the choice $+ + + + +$: if a sequence possesses Property 1, then it cannot possess Property 5).

(a) Find all of the possible choices. For each choice find a sequence corresponding to it.

(b) Prove that the remaining choices are impossible.

**127.** Prove that if a sequence has a limit, then it has either a greatest term or a least term or both. Construct examples of all three cases.

**128.** Prove that any infinite sequence has an infinite monotone subsequence. (The sequence $\{x_n\}$ is said to be monotone if one of the following conditions is satisfied:

1. $x_1 \leq x_2 \leq x_3 \leq \ldots \leq x_n \leq \ldots$,

2. $x_1 \geq x_2 \geq x_3 \geq \ldots \geq x_n \geq \ldots$.

In the first case the sequence is said to be nondecreasing, and in the second case, nonincreasing.)

In the theory of limits there is a property of the real numbers which is very important and is usually accepted as an axiom.

*The Bolzano-Weierstrass Axiom:*

*Every bounded monotone sequence has a limit.*

This axiom reflects the property of "completeness" of the set of real numbers. Figuratively speaking, it expresses the fact that the number axis has no "holes."

In courses in mathematical analysis it is proved that the Bolzano-Weierstrass Axiom is equivalent to each of the following propositions.

1. If an infinite sequence of segments on a number axis has the property that each segment lies inside its predecessor, then the segments have at least one point in common.

2. Every real number can be written in the form of an infinite (periodic or nonperiodic) decimal fraction, and each such fraction corresponds to a real number.

If one of these propositions is taken as an axiom, then the other proposition and the Bolzano-Weierstrass Axiom become provable theorems.

**129*.** Prove that the Bolzano-Weierstrass Axiom is not satisfied by the rational numbers (that is, prove that there exists a bounded monotone sequence of rational numbers which does not have rational limit).

**130.** Prove that every bounded sequence has at least one limit point.

**131.** Prove that the following sequences have limits:

(a) $1, 1 + \frac{1}{4}, 1 + \frac{1}{4} + \frac{1}{9}, \ldots, 1 + \frac{1}{4} + \frac{1}{9} + \ldots + \frac{1}{n^2}, \ldots,$

(b) $1, 1 - \frac{1}{3}, 1 - \frac{1}{3} + \frac{1}{5}, \ldots, 1 - \frac{1}{3} + \frac{1}{5} - \ldots + \frac{(-1)^{n-1}}{2n-1}, \ldots.$

### 3. Problems on the Computation of Limits

**132°.** Prove that if $\lim_{n\to\infty} x_n = a$ and $\lim_{n\to\infty} y_n = b$, then

(a) $\lim_{n\to\infty} (x_n + y_n) = a + b$,

(b) $\lim_{n\to\infty} (x_n - y_n) = a - b$,

(c) $\lim_{n\to\infty} (x_n y_n) = ab$,

(d) $\lim_{n\to\infty} \frac{x_n}{y_n} = \frac{a}{b}$, if $b \neq 0$ and $y_n \neq 0$.

**133°.** Find sequences $\{x_n\}$ and $\{y_n\}$ for which $\lim_{n\to\infty} x_n = 0$, $\lim_{n\to\infty} y_n = 0$, and such that

(a) $\dfrac{x_n}{y_n} \to 0$ as $n \to \infty$,

(b) $\dfrac{x_n}{y^n} \to 1$ as $n \to \infty$,

(c) $\dfrac{x_n}{y_n} \to \infty$ as $n \to \infty$,

(d) $\lim_{n\to\infty} \dfrac{x_n}{y_n}$ does not exist.

**134°.** Find the limits of the following sequences:

(a) $x_n = \dfrac{2n + 1}{3n - 5}$,

(b) $x_n = \dfrac{10n}{n^2 + 1}$,

(c) $x_n = \dfrac{n(n + 2)}{(n + 1)(n + 3)}$,

(d) $x_n = \dfrac{2^n + 1}{2^n - 1}$,

(e) $x_n = \dfrac{1}{n^k + 1}(1^k + 2^k + \ldots + n^k)$
($k$ is a fixed natural number).

The sequence in Problem 134e has the following geometrical interpretation. Consider the part of the plane bounded by the graph of the function $y = x^k$, the $x$-axis, and the line $x = 1$. Divide the segment $[0, 1]$ on the $x$-axis into $n$ equal parts and construct a rectangle on each part in such a way that the upper right-hand vertex lies on the graph of our function (Fig. 17). The sum of the areas of all the rectangles which we have constructed is immediately seen to be equal to the quantity $x_n = 1/n^{k+1}(1^k + 2^k + \ldots + n^k)$. The limit of this quantity as $n \to \infty$ can be defined to be the area of the curved figure under consideration.

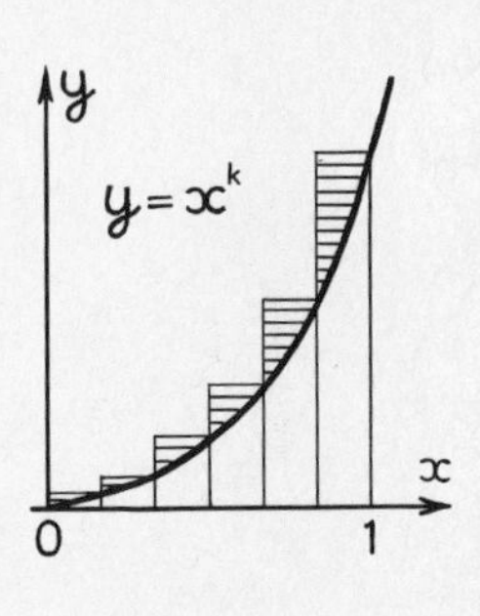

Fig. 17

**135°.** Prove that $\lim_{n\to\infty} (\sqrt{n+1} - \sqrt{n-1}) = 0$.

**136.** Prove that $\lim_{n\to\infty} n/2^n = 0$.

**137.** Prove that $\lim_{n\to\infty} n/a^n = 0$ for $a > 1$.

**138*.** Prove that $\lim_{n\to\infty} \dfrac{\log_2 n}{n} = 0$.

**139*.** Find $\lim_{n\to\infty} \sqrt[n]{n}$.

Until now we have spoken only of the limits of sequences of numbers. But by means of numbers we can specify various geometrical objects. The direction of a straight line on a plane, for example, can be specified by its slope; a point on a line or on a plane can be specified by its coordinates; and so on. Whenever the term "limit" or "convergence" is applied to a sequence of geometric objects, we are really talking about a numerical sequence characterizing these objects. Thus the statement "The sequence of points $M_n$ in the plane converges to the point $M$" means that the coordinates of the points $M_n$ converge to the corresponding coordinates of the point $M$.

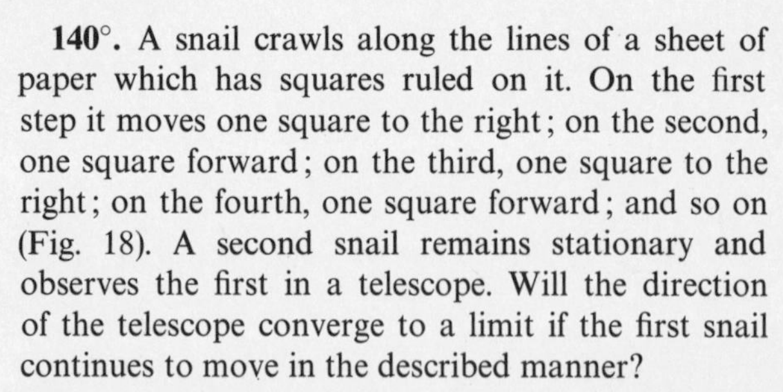

**140°.** A snail crawls along the lines of a sheet of paper which has squares ruled on it. On the first step it moves one square to the right; on the second, one square forward; on the third, one square to the right; on the fourth, one square forward; and so on (Fig. 18). A second snail remains stationary and observes the first in a telescope. Will the direction of the telescope converge to a limit if the first snail continues to move in the described manner?

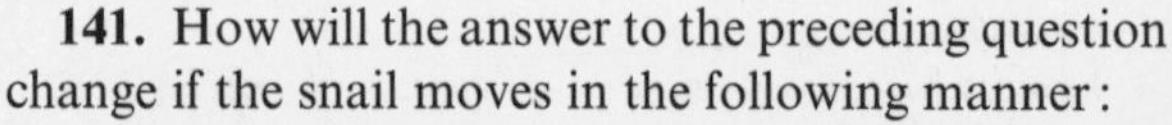

**141.** How will the answer to the preceding question change if the snail moves in the following manner:

(a) 1 square to the right, 2 squares forward; 1 square to the right; 2 squares forward; and so on?

(b) 1 square to the right, 2 squares forward; 3 squares to the right, 4 squares forward; 5 squares to the right, 6 squares forward; and so on?

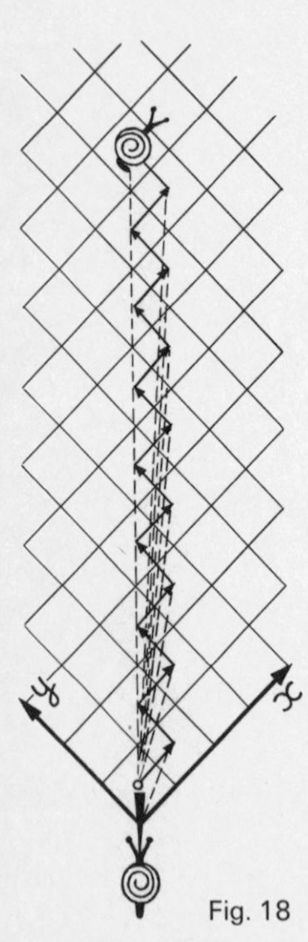


Fig. 18

(c) 1 square to the right, 2 squares forward; 4 squares to the right, 8 squares forward; 16 squares to the right, 32 squares forward; and so on (Fig. 19)?

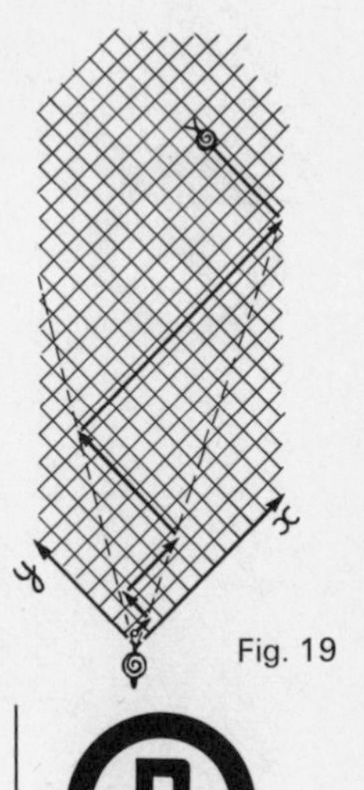
Fig. 19

**142°.** On the parabola that is the graph of the function $y = x^2$ take the point $A_0$ with abscissa $a$ and the sequence of points $A_n$ with abscissa $a + \frac{1}{n}$. Denote by $M_n$ the point of intersection of the $x$-axis with the secant drawn through the points $A_0$ and $A_n$. Prove that the sequence of points $M_n$ has a limit $M_0$ as $n \to \infty$ and find this limit.

The straight line $A_0M_0$ is called the *tangent to the parabola at the point* $A_0$.

**143°.** Little Peter walked out of his house and started on the way to school. Having gone halfway to school, he decided that it would be better to go to the movies and turned toward the movie theater. When he had traveled half of the distance, he decided that he would rather go skating. Having gone half of the distance to the skating rink, he suddenly realized that he had better study after all and turned toward school. But when he was halfway to school he again turned toward the movie theater (Fig. 20). Where did Peter wind up if he continued in this way?

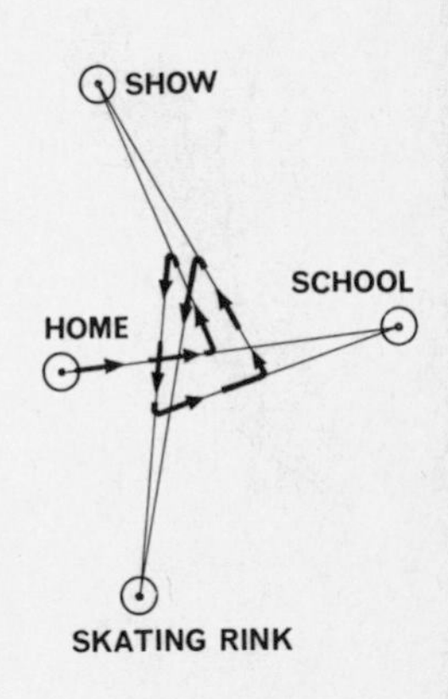

Fig. 20

**144°.** On a line, construct the sequence of points $\{M_n\}$ according to the following rule. Choose the first two points $M_1$ and $M_2$ arbitrarily, and take each point after this to be the midpoint of the segment joining the two preceding points. Prove that the limit of the sequence $\{M_n\}$ exists and find it.

**145.** Define the *sum of an infinite series* of numbers in the following manner. Given the series

$$a_1 + a_2 + a_3 + \ldots + a_n + \ldots,$$

denote by $S_n$ the sum of the first $n$ terms. If the sequence $\{S_n\}$ has a limit $S$, we call the number $S$ the sum of the given series. If the sequence $\{S_n\}$ does not have a limit, we say that the given series diverges and we do not assign it a sum.

Find the sums

(a) $1 + a + a^2 + \ldots + a^2 + \ldots,$

(b) $a + 2a^2 + 3a^3 + \ldots + na^n + \ldots,$

(c) $\dfrac{1}{1\cdot 2} + \dfrac{1}{2\cdot 3} + \dfrac{1}{3\cdot 4} + \ldots + \dfrac{1}{n(n+1)} + \ldots,$

(d) $\dfrac{1}{1\cdot 2\cdot 3} + \dfrac{1}{2\cdot 3\cdot 4} + \dfrac{1}{3\cdot 4\cdot 5} + \ldots + \dfrac{1}{n(n+1)(n+2)} + \ldots,$

(e)* $1 + \frac{1}{2} + \frac{1}{3} + \ldots + \frac{1}{n} + \ldots$

**146.** Suppose that one has an infinite collection of bricks that are in the form of rectangular parallelepipeds. The bricks are placed on top of one another with a certain displacement so that they do not fall (Fig. 21). How long a "roof" can be constructed in this way?

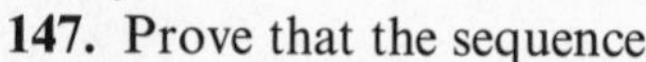

Fig. 21

**147.** Prove that the sequence

$$2,\ 2 + \tfrac{1}{2},\ 2 + \frac{1}{2 + \frac{1}{2}},\ 2 + \cfrac{1}{2 + \cfrac{1}{2 + \frac{1}{2}}}, \ldots$$

has a limit and find it.

**148.** For computing the square root of a positive number $a$ one can use the following method of successive approximation. Take an arbitrary number $x_0$ and construct a sequence according to the following rule:

$$x_{n+1} = \frac{1}{2}\left(x_n + \frac{a}{x_n}\right).$$

Prove that if $x_0 > 0$, then $\lim_{n \to \infty} x_n = \sqrt{a}$, whereas if $x_0 < 0$, $\lim_{n \to \infty} x_n = \sqrt{-a}$. (The sign $\sqrt{a}$ denotes the square root of $a$.)

How many successive approximations are necessary (that is, how many terms of the sequence $\{x_n\}$ is it necessary to compute), in order to find the value of $\sqrt{10}$ to an accuracy of 0.00001 if one takes $x_0 = 3$ as an initial value?

# Test Problems

## For Chapter 1

**1.** Compute the sum

$$\frac{1}{1\cdot 3}+\frac{1}{3\cdot 5}+\ldots+\frac{1}{(2n-1)(2n+1)}.$$

**2.** Prove that for any natural number $n$,

$$\frac{1^2}{1\cdot 3}+\frac{2^2}{3\cdot 5}+\ldots+\frac{n^2}{(2n-1)(2n+1)}=\frac{n(n+1)}{2(2n+1)}.$$

**3.** Find the product

$$\left(1-\frac{4}{9}\right)\left(1-\frac{4}{16}\right)\left(1-\frac{4}{25}\right)\ldots\left(1-\frac{4}{n^2}\right),$$

where $n \geq 3$.

**4.** Prove that

$$\frac{(n+1)(n+2)\ldots(2n-1)2n}{1\cdot 3\cdot 5\ldots(2n-1)}=2^n.$$

**5.** Prove that for any natural number $n > 1$ the following inequality is valid:

$$\frac{1}{(n+1)}+\frac{1}{(n+2)}+\ldots+\frac{1}{2n}>\frac{13}{24}.$$

**6.** For what natural numbers $n$ is the following inequality valid:

$$2^n > 2n + 1?$$

**7.** Prove that

$$2^{n-1}(a^n + b^n) > (a + b)^n,$$

where $a + b > 0$, $a \neq b$, $n > 1$.

**8.** Prove that a plane divided into $n$ parts by circles can be painted using black and white paint so that any two adjacent parts will be different colors.

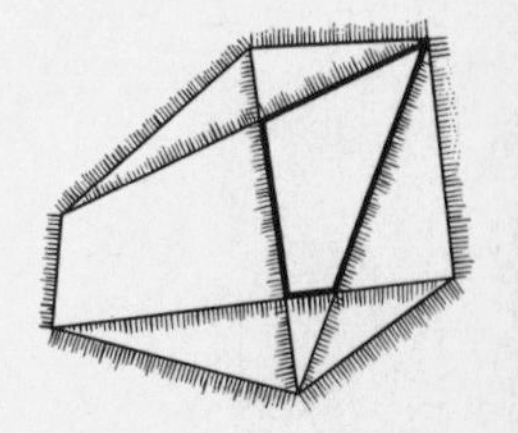

Fig. 22

**9.** Paint the exterior of a convex polygon. Construct several diagonals of the polygon in such a way that no three of them intersect in a point. Paint one side of each of these diagonals, that is, draw a narrow colored strip on one side (Fig. 22). Prove that you must have painted the entire exterior of at least one of the polygons into which the diagonals partition the initial polygon.

**10.** Prove that for every integer $n \geq 0$ the number $10^{n+1} - 10(n + 1) + n$ is divisible by 81.

**11.** Find the sum

$$1 \cdot 3 + 3 \cdot 5 + \ldots + (2n - 1)(2n + 1).$$

**12.** In an arithmetic progression the sum of the first $n$ terms is equal to the sum of the first $m$ terms, $n \neq m$. Prove that the sum of the first $n + m$ terms of this progression is equal to 0.

**13.** Find the sum

$$6 + 66 + 666 + \ldots + \underbrace{666 \ldots 66}_{n \text{ times}}.$$

**14.** Does there exist an arithmetic progression among the terms of which are the numbers, 1, $\sqrt{2}$ and 3? These numbers need not be consecutive nor appear in the preceding order.

**15.** Given a geometric progression the ratio of which is a whole number not equal to 0 or $-1$, prove that the sum of any number of arbitrarily chosen terms cannot equal any term of this progression.

**16.** Each side of a right triangle is divided into $n$ equal parts. Through each point of the partition a straight line is constructed parallel to the sides. These lines partition the triangle into equal, smaller triangles. Some of these triangles are painted black, some white, so that any black triangle is bounded by an even number of white ones and any white one is bounded by an odd number of white ones. Prove that the small triangles located in the corners of the large one are all the same color.

**17.** Find the error in the following "proof" by induction of the proposition "All numbers are equal."

**Proof.** We prove that any $n$ numbers are equal. One number is equal to itself. Therefore the proposition is true for $n = 1$. Suppose that it is true for $n = k$; we prove that it is true for $n = k + 1$.

Let us enumerate all of the numbers of the given set by numbers $1, 2, 3, \ldots, k + 1$. The first $k$ numbers are all equal and therefore are equal to the first number. Exclude the second number. Then the remaining $k$ numbers, among which is the $(k + 1)$st number, are all equal and are equal to the first number.

Thus, all of the numbers are equal to the first; and thus are all equal. This completes the proof.

## For Chapter 2

**18.** The numbers 1 to 6 are placed on the faces of a cube. We thus get a die. Find the number of different dice.

**19.** Which is more likely to occur as the sum in a double toss of a die: 9 or 10?

**20.** At the end of a game of dominoes all of the pieces happen to be laid out in a chain. Find the number of possible chains (the pieces are arranged in a straight line).

**21.** (a) How many ways are there of seating 19 people around a circular table?

(b) How many ways are there of seating 19 people around a circular table in such a way that exactly $r$ people are sitting between $A$ and $B$?

**22.** Two soccer teams play "to 10" The judges write down on the scoresheet how the score changes. For example: 1:0, 1:1, 1:2, 1:3, 1:4, 2:4, 2:5, 2:6, 3:6, 4:6, 5:6, 6:6, 7:6, 8:6, 9:6, 9:7, 10:7. How many different score sheets can be obtained?

**23.** A child's rattle consists of a ring with three white beads and seven red ones strung on it. Some rattles, seemingly different, can be made identical by arranging the rings and moving the beads in a suitable manner (Fig. 23). Find the number of essentially different rattles.

**24.** In how many ways can a piece get from square $a1$ on a chessboard to square $h8$ if it can move only one square forward or one square to the right at a time?

**25.** In how many ways can one place two white rooks and two black ones on a chessboard so that the white pieces do not threaten the black ones?

**26.** In how many ways can a checker on $a1$ be crowned? (There are no other checkers on the board.)

**27.** By how many non-self-intersecting paths can one get from $A$ to $B$ (Fig. 24)? (The movement is to be along a circular and radial route with transfers at the positions marked off with white circles.)

**28.** In how many ways can one place 9 rooks on a three-dimensional $3 \times 3 \times 3$ chessboard so that they do not threaten one another? (A rook threatens its row, its column, and its file.)

**29.** A number from 1 to $n$ is entered in each square of a $2 \times n$ table in such a way that no number appears twice in any one row or column. How many such tables are there?

**30.** A forgetful passenger left his things in an automatic locker, and one hour before the train's departure, he discovered that he had almost forgotten the number. He remembered only that 23 and 37 occurred in the number. How many numbers did he have to

Fig. 23

Fig. 24

sort over? (In order to get his belongings, he had to choose correctly a number of five digits.)

**31.** A man had 7 daughters. Whenever one of them married, each unmarried older sister went to her father to complain that the old custom of marrying in order of age had been broken. By the time the last daughter had married, the father had heard 7 complaints. In how many ways could this have happened?

A more serious formulation of this problem is the following. Let us call any arrangement of the sequence of the first $n$ natural numbers in which each number occurs once an *n-permutation*. We shall call a pair of numbers $(p, l)$ of this sequence an *irregularity* if $p > l$ and $p$ is farther to the left in the $n$-permutation than $l$. How many $n$-permutations are there with $k$ irregularities? (In the problem concerning the sisters $n = k = 7$.)

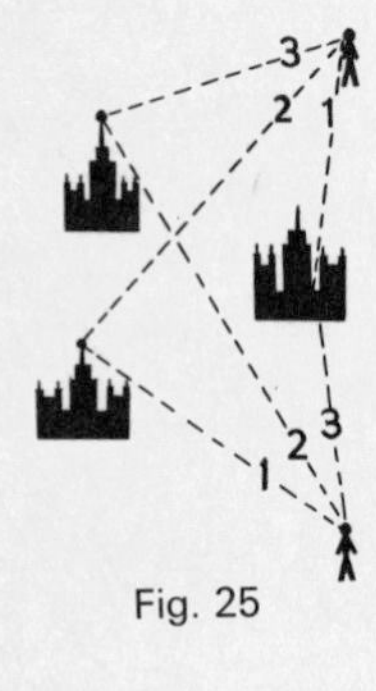

Fig. 25

**32.** If there are two tall buildings in a city, their spires will be visible from somewhere in the city in either order (the spire of the first building to the left, that of the second to the right, and vice versa). The same thing is true for three tall buildings provided that they do not form a straight line (Fig. 25). An architect wishes to lay out 7 tall buildings in such a way that for each possible ordering of their spires from left to right at some point in the city their spires will appear in that order. Will he be able to do this?

**For Chapter 3**

**33.** Prove that the number 2 is not the limit of the sequence

$1, 3, \frac{1}{2}, 2\frac{1}{2}, \frac{1}{3}, 2\frac{1}{3}, \ldots, \frac{1}{n}, 2\frac{1}{n}, \ldots.$

**34.** Given that $\lim_{n\to\infty} x_n = 1$, find the limits of the following sequences:

(a) $y_n = \dfrac{2x_n - 1}{x_n + 1}$ (b) $y_n = \dfrac{x_n^2 + x_n - 2}{x_n - 1}$

(c) $y_n = \dfrac{x_n^{10} - 1}{x_n - 1}$ (d) $y_n = \sqrt{x_n}$.

**35.** Prove that if the sequence $\{x_n\}$ has the limit $a$, then the sequence obtained by any rearrangement of the terms also has the limit $a$.

**36.** Prove that if a sequence has the limit $a$, then any infinite subsequence also has the limit $a$.

**37.** Prove that if the number $a$ is a limit point of the sequence $\{x_n\}$, then one can choose a subsequence of $\{x_n\}$ having $a$ as a limit. Is the converse true?

**38.** Find a sequence for which every limit point is a whole number and each whole number is a limit point.

**39.** Prove that if $x_n \geq 0$ for all $n$ and $\lim\limits_{n\to\infty} x_n = a$, then $a \geq 0$.

**40.** Find the limits:

(a) $\lim\limits_{n\to\infty} \dfrac{n^2 + 3n - 2}{1 + 2 + 3 + \ldots + n}$

(b) $\lim\limits_{n\to\infty} \dfrac{n^2}{4^n}$

(c) $\lim\limits_{n\to\infty} \dfrac{n + \log n + 2^n}{n^2 - \log n - 2^n}$

(d) $\lim\limits_{n\to\infty} (\sqrt{n^2 + n} - \sqrt{n^2 - n})$.

**41.** The sequence $\{x_n\}$ is constructed according to the following rule: the first term is chosen arbitrarily and each following one is chosen according to the formula $x_{n+1} = ax_n + b$, where $a$ and $b$ are constants. For what $a$ and $b$ does the sequence $\{x_n\}$ have a limit?

**42*.** Prove that if in the series

$$1 + \tfrac{1}{2} + \tfrac{1}{3} + \tfrac{1}{4} + \ldots + \tfrac{1}{n} + \ldots$$

one removes all of the terms in whose denominator the digit 3 appears, then the sum of the series of the remaining terms will be finite.

**43.** On the graph of the function $y = x^2$ consider the points $A_n$ and $B_n$ with abscissas $1/n$ and $-1/n$, respectively. Draw a circle through $A_n$, $B_n$, and the origin. Let $M_n$ be the center of this circle (Fig. 26). Prove that the sequence of points $\{M_n\}$ has a limit and find it.

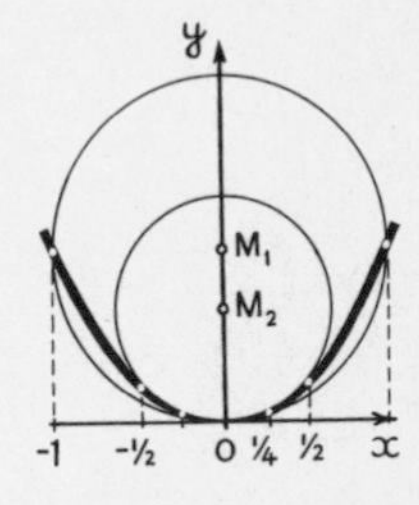

Fig. 26

# Solutions

## Chapter 1

**1.** We solve the problem by the method of mathematical induction.

(a) For $n = 1$ the equality is valid.

(b) Suppose that the equality is valid for $n = k$, where $k$ is any integer. We shall prove that in such a case it is also valid for $n = k + 1$, that is, that

$$1^3 + 2^3 + 3^3 + \ldots + k^3 + (k+1)^3 = \frac{(k+1)^2(k+2)^2}{4}.$$

For suppose that for $n = k$ the equality is valid; that is, suppose that

$$1^3 + 2^3 + 3^3 + \ldots + k^3 = \frac{k^2(k+1)^2}{4}.$$

Then

$$1^3 + 2^3 + 3^3 + \ldots + k^3 + (k+1)^3 = \frac{k^2(k+1)^2}{4} + (k+1)^3 = (k+1)^2\left[\frac{k^2}{4} + k + 1\right] = \frac{(k+1)^2}{4}(k^2 + 4k + 4) = \frac{(k+1)^2(k+2)^2}{4}.$$

Thus the validity of the equality for $n = k$ implies its validity for $n = k + 1$. Therefore, by the principle of mathematical induction we can assert that for all natural numbers $n$ the equality

$$1^3 + 2^3 + 3^3 + \ldots + n^3 = \frac{n^2(n+1)^2}{4}$$

is valid.

**2. First Solution.** We apply the principle of mathematical induction.

(a) We verify the validity of the equality for $n = 1$ directly.

(b) Assume that the equality is valid for a natural number $k$, that is, that

$$\frac{1}{1\cdot 2} + \frac{1}{2\cdot 3} + \frac{1}{3\cdot 4} + \ldots + \frac{1}{k(k+1)} = \frac{k}{k+1}.$$

We prove that it is then valid for $n = k + 1$ as well. In fact

$$\frac{1}{1\cdot 2} + \frac{1}{2\cdot 3} + \frac{1}{3\cdot 4} + \ldots + \frac{1}{k(k+1)} + \frac{1}{(k+1)(k+2)}$$

$$= \frac{k}{k+1} + \frac{1}{(k+1)(k+2)}$$

$$= \frac{k(k+2)+1}{(k+1)(k+2)} = \frac{k^2+2k+1}{(k+1)(k+2)}$$

$$= \frac{(k+1)^2}{(k+1)(k+2)} = \frac{k+1}{k+2}.$$

Thus we have proved both of the assertions that must be valid for the method of mathematical induction to be applicable. Therefore our equality is true for all $n$.

**Second Solution.** For any $n$ the following formula holds:

$$\frac{1}{n(n+1)} = \frac{1}{n} - \frac{1}{n+1}$$

(this can be verified directly). Therefore the left-hand side of the equality that we want to prove can be written in the following form:

$$\frac{1}{1\cdot 2} + \frac{1}{2\cdot 3} + \frac{1}{3\cdot 4} + \ldots + \frac{1}{(n-1)n} + \frac{1}{n(n+1)}$$

$$= \left(\frac{1}{1} - \frac{1}{2}\right) + \left(\frac{1}{2} - \frac{1}{3}\right) + \left(\frac{1}{3} - \frac{1}{4}\right) + \ldots +$$

$$+ \left(\frac{1}{n-1} - \frac{1}{n}\right) + \left(\frac{1}{n} - \frac{1}{n+1}\right)$$

$$= 1 - \frac{1}{2} + \frac{1}{2} - \frac{1}{3} + \frac{1}{3} - \frac{1}{4} + \ldots +$$

$$+ \frac{1}{n-1} - \frac{1}{n} + \frac{1}{n} - \frac{1}{n+1}.$$

In the sum obtained all of the summands except the first and the last cancel one another. Therefore

$$\frac{1}{1\cdot 2} + \frac{1}{2\cdot 3} + \frac{1}{3\cdot 4} + \ldots + \frac{1}{(n-1)n} + \frac{1}{n(n+1)}$$

$$= 1 - \frac{1}{n+1} = \frac{n}{n+1},$$

which is what we wanted to prove.

**3. First Solution.** We prove by the method of mathematical induction that the equation

$$\frac{1}{1\cdot 4} + \frac{1}{4\cdot 7} + \ldots + \frac{1}{(3n-2)(3n+1)} = \frac{n}{3n+1}$$

is valid.

(a) For $n = 1$ the equality is valid.

(b) Suppose that this equation is satisfied for a natural number $k$; that is, suppose that

$$\frac{1}{1\cdot 4} + \frac{1}{4\cdot 7} + \ldots + \frac{1}{(3k-2)(3k+1)} = \frac{k}{3k+1}.$$

We shall show that it will then be valid for $n = k + 1$ as well.

In fact,

$$\frac{1}{1\cdot 4} + \frac{1}{4\cdot 7} + \ldots + \frac{1}{(3k-2)(3k+1)} +$$

$$+ \frac{1}{(3k+1)(3k+4)} = \frac{k}{3k+1} + \frac{1}{(3k+1)(3k+4)}$$

$$= \frac{3k^2+4k+1}{(3k+1)(3k+4)} = \frac{(3k+1)(k+1)}{(3k+1)(3k+4)} = \frac{k+1}{3k+4}.$$

Thus, the validity of the equation is proved for all $n$.

**Second Solution.** We use the formula

$$\frac{1}{(3n-2)(3n+1)} = \frac{1}{3(3n-2)} + \frac{1}{3(3n+1)}.$$

Applying this to all of the terms in our sum, we get

$$\frac{1}{1\cdot 4} + \frac{1}{4\cdot 7} + \frac{1}{7\cdot 10} + \ldots + \frac{1}{(3n-5)(3n-2)} +$$

$$+ \frac{1}{(3n-2)(3n+1)} = \left(\frac{1}{3\cdot 1} - \frac{1}{3\cdot 4}\right) +$$

$$+ \left(\frac{1}{3\cdot 4} - \frac{1}{3\cdot 7}\right) + \left(\frac{1}{3\cdot 7} - \frac{1}{3\cdot 10}\right) + \ldots +$$

$$+ \left(\frac{1}{3(3n-5)} - \frac{1}{3(3n-2)}\right) +$$

$$+ \left(\frac{1}{3(3n-2)} - \frac{1}{3(3n+1)}\right).$$

Upon removal of the parentheses, all of the summands except the first and last are canceled; and so we get

$$\frac{1}{1\cdot 4} + \frac{1}{4\cdot 7} + \frac{1}{7\cdot 10} + \ldots + \frac{1}{(3n-2)(3n+1)}$$

$$= \frac{1}{3} - \frac{1}{3(3n+1)} = \frac{n}{3n+1}.$$

The equation is thus proved.

**4.** We show here a solution based on the principle of mathematical induction.

(a) By a direct verification we easily convince ourselves that for $n = 1$ the equation is valid.

(b) Suppose that the equation is valid for $n = k$:

$$\frac{1}{a(a+1)} + \frac{1}{(a+1)(a+2)} + \ldots + \frac{1}{(a+k-1)(a+k)} = \frac{k}{a(a+k)}.$$

We show that it is then valid for $n = k + 1$ as well. We have

$$\begin{aligned}
&\frac{1}{a(a+1)} + \frac{1}{(a+1)(a+2)} + \ldots + \\
&\quad + \frac{1}{(a+k-1)(a+k)} + \frac{1}{(a+k)(a+k+1)} \\
&\quad = \frac{k(a+k+1)+a}{a(a+k)(a+k+1)} = \frac{k^2+ak+a+k}{a(a+k)(a+k+1)} \\
&\quad = \frac{(a+k)(k+1)}{a(a+k)(a+k+1)} = \frac{k+1}{a(a+k+1)}.
\end{aligned}$$

Therefore, by the principle of mathematical induction, we can assert that the equation is valid for all natural numbers $n$.

**5.** We solve the problem by the method of mathematical induction.

(a) For $n = 1$ the equation is valid (Check this!).

(b) Suppose that the equation is valid for $n = k$:

$$1 \cdot 1! + 2 \cdot 2! + \ldots + k \cdot k! \quad (k+1)! - 1.$$

We show that in this case it will be satisfied for $n = k + 1$. In fact,

$$\begin{aligned}
&1 \cdot 1! + 2 \cdot 2! + \ldots + k \cdot k! + (k+1)(k+1)! \\
&\quad = (k+1)! - 1 + (k+1)(k+1)! \\
&\quad = (k+2)(k+1)! - 1.
\end{aligned}$$

But from the definition of $n!$ it is clear that

$$(k+2) \cdot (k+1)! = (k+2)!.$$

Therefore

$$1 \cdot 1! + 2 \cdot 2! + \ldots + (k + 1)(k + 1)! = (k + 2)! - 1;$$

that is, our equation turns out to be true for $n = k + 1$. Applying the method of mathematical induction, we find that our equation is valid for all natural numbers $n$.

**6. First Solution.** We shall prove our equation by the method of mathematical induction.

(a) For $n = 2$ a direct verification shows the validity of the equation.

(b) Suppose that the equation is valid for a natural number $k \geq 2$:

$$\left(1 - \frac{1}{4}\right)\left(1 - \frac{1}{9}\right)\left(1 - \frac{1}{16}\right)\ldots\left(1 - \frac{1}{k^2}\right) = \frac{k + 1}{2k}.$$

We shall show that in this case it is valid also for $n = k + 1$. In fact,

$$\left(1 - \frac{1}{4}\right)\left(1 - \frac{1}{9}\right)\left(1 - \frac{1}{16}\right)$$

$$\ldots\left(1 - \frac{1}{k^2}\right)\left(1 - \frac{1}{(k + 1)^2}\right) = \frac{k + 1}{2k}\left(1 - \frac{1}{(k + 1)^2}\right)$$

$$= \frac{(k + 1)[(k + 1)^2 - 1]}{2k(k + 1)^2} = \frac{(k + 1)k(k + 2)}{2k(k + 1)^2}$$

$$= \frac{k + 2}{2(k + 1)}.$$

Applying the principle of mathematical induction, we can now assert that the equation is valid for all $n \geq 2$.

**Second Solution.** For each natural number $n \geq 2$ the following formula is valid:

$$1 - \frac{1}{n^2} = \frac{(n - 1)(n + 1)}{n^2}.$$

Applying this formula to each factor in the left side of the equation, we obtain:

$$\left(1-\frac{1}{4}\right)\left(1-\frac{1}{9}\right)\left(1-\frac{1}{16}\right)$$

$$\ldots\left(1-\frac{1}{(n-1)^2}\right)\left(1-\frac{1}{n^2}\right)$$

$$=\frac{(2-1)(2+1)}{2^2}\cdot\frac{(3-1)(3+1)}{3^2}\cdot\frac{(4-1)(4+1)}{4^2}$$

$$\ldots\frac{(n-2)n}{(n-1)^2}\cdot\frac{(n-1)(n+1)}{n^2}.$$

It is easy to see that almost all of the factors of the numerator and the denominator cancel one another. Only $2 - 1 = 1$ and $n + 1$ in the numerator and 2 and $n$ in the denominator fail to be canceled. Therefore,

$$\left(1-\frac{1}{4}\right)\left(1-\frac{1}{9}\right)\left(1-\frac{1}{16}\right)\ldots\left(1-\frac{1}{n^2}\right)=\frac{n+1}{2n};$$

that is, the equation is proved.

**7.** We apply the method of mathematical induction.

(a) For $n = 1$ the formula is valid (check this!).

(b) Suppose that the formula is valid for some natural number $k$, that is, suppose that

$$1-2^2+3^2\ldots+(-1)^{k-1}k^2=(-1)^{k-1}\frac{k(k+1)}{2}.$$

We shall show that it is then valid for $n = k + 1$. In fact,

$$1-2^2+3^2-\ldots+(-1)^{k-1}k^2+(-1)^k(k+1)^2$$

$$=(-1)^{k-1}\frac{k(k+1)}{2}+(-1)^k(k+1)^2$$

$$=(-1)^{k-1}(k+1)\left(\frac{k}{2}-k-1\right)$$

$$=(-1)^{k-1}\frac{k+1}{2}(-k-2)=(-1)^k\frac{(k+1)(k+2)}{2}.$$

Therefore, using the principle of mathematical induction, we get the validity of our formula for all natural numbers $n$.

**8.** Suppose that $n$ lines have already been drawn. If another line is drawn it will intersect the first $n$ lines in $n$ points, for by our condition it will intersect each of them and will not intersect any two in the same point. It is clear that these $n$ points will divide the $(n + 1)$st line into $n + 1$ segments (two of them infinite). Each of these segments will cut one of the regions defined by the first $n$ lines and thereby define two new regions. Therefore the $n + 1$ segments together will give rise to $2n + 2$ new regions of the plane from the $n + 1$ old regions, that is, the number of regions of the plane will increase by $n + 1$ upon the addition of the $(n + 1)$st line. Now we can easily compute the number of regions into which $n$ straight lines divide the plane. One straight line divides the plane into 2 regions. The addition of the second line increases the number of regions by 2; the addition of the third by 3; and so on. Thus the total number of regions is equal to

$$2 + 2 + 3 + 4 + \ldots + n$$

$$= 1 + (1 + 2 + 3 + \ldots + n) = 1 + \frac{n(n + 1)}{2}$$

(See earlier, Example 9 on page 4. The number of regions into which the lines divide the plane is thus equal to $[n(n + 1)]/2 + 1$.

**9.** (a) For $n = 8$ the assertion is true, since 8 kopeks can be represented by three- and five-kopek coins: $8 = 5 + 3$.

(b) Suppose that we can represent $k$ kopeks in the form of a sum of three- and five-kopek coins. We shall show that it will then be possible to do the same thing for $k + 1$ kopeks. We consider the case where the representation of $k$ kopeks involves at least one five-kopek coin and the case where the representation involves only three-kopek coins. In the first case we get the required representation by exchanging one five-kopek coin for two three-kopek ones. In the second case it is clear that the number of three-kopek coins is not less than three. Therefore we can exchange three three-kopek coins for two five-kopek ones and thereby get the required representation of $k + 1$ kopeks. Thus if $k$ kopeks can be represented as a sum of three- and five-kopek coins, then the same thing can be done for $k + 1$ kopeks. Using the principle of mathematical induction, we therefore find that any amount of money in excess of 7 kopeks can be represented by three- and five-kopek coins.

**10.** (a) If the plane is cut by only one line, it is possible to paint it as required. It suffices merely to paint one half-plane one color, and the second the other color.

(b) Suppose that a plane upon which $n$ straight lines are drawn has been painted in the required manner. We draw the $(n + 1)$st line and prove that the plane that is obtained can also be painted in the required manner. The new line divides the plane into two half-planes. Therefore, in order to obtain the necessary design, we proceed as follows. In one of the half-planes that we have obtained we change the color of each piece (that is, we change white to black and black to white), but we leave the other half-plane unchanged. We claim that the plane will now be properly colored. For consider two neighboring regions. Two cases are possible: the line dividing them is one of the original lines, or it is the $(n + 1)$st line. In the first case the regions were painted different colors until the new line was drawn and the colors changed, and then either they both changed color or they both remained unchanged. Clearly then, they still are of different color. In the second case our regions were the same color until the new line was drawn, and then one of the regions changed color; thus, the two regions will be painted different colors. Therefore, if we have painted the plane divided into parts by $n$ lines correctly, then we can paint the plane correctly if it is divided into regions by $n + 1$ lines. We can now apply the principle of mathematical induction which proves the validity of our assertion.

**11.** It is necessary to prove that for any natural number $n$ the number $n^3 + (n + 1)^3 + (n + 2)^3$ is divisible by 9.

(a) For $n = 1$ the assertion is true, because

$$1^3 + 2^3 + 3^3 = 36$$

is divisible by 9.

(b) Suppose that the assertion is valid for a natural number $k$, that is, suppose that $k^3 + (k + 1)^3 + (k + 2)^3$ is divisible by 9. We shall prove that the assertion is then valid for $n = k + 1$ as well. We have

$$\begin{aligned}(k + 1)^3 + (k + 2)^3 + (k + 3)^3 &= (k + 1)^3 + \\ &+ (k + 2)^3 + (k^3 + 3\cdot 3k^2 + 3\cdot 3^2\cdot k + 3^3) \\ &= [k^3 + (k + 1)^3 + (k + 2)^3] + 9(k^2 + 3k + 3).\end{aligned}$$

We have represented the number $(k + 1)^3 + (k + 2)^3 + (k + 3)^3$ as a sum of two terms, of which the first is divisible by 9 by assumption, and the second is the product of 9 and a whole number. Thus the sum of these terms is divisible by 9.

We can thus use the principle of mathematical induction, which proves our assertion.

**12.** (a) Let us check our assertion for $n = 0$. In this case

$$11^{n+2} + 12^{2n+1} = 11^2 + 12 = 121 + 12 = 133$$

is divisible by 133.

(b) Suppose that our proposition is satisfied for some $k \geq 0$, that is, suppose that $11^{k+2} + 12^{2k+1}$ is divisible by 133. We shall show that in this case it is satisfied for $n = k + 1$ as well. We have

$$\begin{aligned}11^{(k+1)+2} + 12^{2(k+1)+1} &= 11^{k+3} + 12^{2k+3} \\ &= 11\cdot 11^{k+2} + 144\cdot 12^{2k+1} = 11\cdot 11^{k+2} + \\ &+ 11\cdot 12^{2k+1} + 133\cdot 12^{2k+1} \\ &= 11(11^{k+2} + 12^{2k+1}) + 133\cdot 12^{2k+1}.\end{aligned}$$

The first summand is divisible by 133 by assumption, and the second contains the number 133 as a factor. Thus the sum is divisible by 133.

The principle of mathematical induction now shows that the assertion is valid for all $n \geq 0$.

**13.** (a) For $n = 2$ the inequality is valid because $(1 + a)^2 = 1 + 2a + a^2 > 1 + 2a$, as $a^2 > 0$ for $a \neq 0$.

(b) Suppose that the inequality is valid for some natural number $k > 2$, that is, suppose that

$$(1 + a)^k > 1 + ka.$$

We shall prove that it is valid also for $n = k + 1$. In fact, by assumption, $1 + a > 0$. Therefore, multiplying both sides of the inequality $(1 + a)^k > 1 + ka$ by $1 + a$, we get

$$(1 + a)^k(1 + a) > (1 + ka)(1 + a),$$
$$(1 + a)^{k+1} > (1 + (k + 1)a) + ka^2.$$

But $ka^2 > 0$ for $a \neq 0$, and therefore

$$(1 + a)^{k+1} > 1 + (k + 1)a.$$

Thus, applying the principle of mathematical induction, we find that for all $n \geq 2$

$$(1 + a)^n > 1 + na,$$

if $a > -1, a \neq 0$.

**14.** (a) For $n = 2$ the inequality is satisfied, since $1 + (1/\sqrt{2}) > \sqrt{2}$.

(b) Suppose that the inequality is satisfied for $n = k$, that is, suppose that

$$1 + \frac{1}{\sqrt{2}} + \frac{1}{\sqrt{3}} + \ldots + \frac{1}{\sqrt{k}} > \sqrt{k}.$$

We shall show that the inequality is then satisfied for $n = k + 1$ as well. For this, it suffices to prove the validity of the inequality

$$\frac{1}{\sqrt{k + 1}} > \sqrt{k + 1} - \sqrt{k}.$$

In fact, if the latter inequality can be proved, then from this inequality and our inequality for $n = k$, which is valid by assumption, we will have our inequality for $n = k + 1$. Let us now prove the inequality

$$\frac{1}{\sqrt{k+1}} > \sqrt{k+1} - \sqrt{k}.$$

We multiply both sides by $\sqrt{k+1} + \sqrt{k}$ to obtain the inequality

$$1 + \sqrt{\frac{k}{k+1}} > 1,$$

which is clearly true. Therefore inequality (*) is also true. Thus, from the validity of the inequality

$$1 + \frac{1}{\sqrt{2}} + \frac{1}{\sqrt{3}} + \ldots + \frac{1}{\sqrt{n}} > \sqrt{n}$$

for $n = k$, its validity for $n = k + 1$ follows.

We now apply the principle of mathematical induction and see that our inequality is valid for all $n \geq 2$.

**15.** (a) For $n = 1$ and $n = 2$ the formula is clearly valid.

(b) Suppose that the formula is valid for all $n \leq k$, where $k \geq 2$. We shall show that it is then valid for $n = k + 1$ as well. In fact,

$$\begin{aligned} u_{k+1} &= 3u_k - 2u_{k-1} = 3(2^{k-1} + 1) - 2(2^{k-2} + 1) \\ &= 3 \cdot 2^{k-1} + 3 - 2^{k-1} - 2 = 2 \cdot 2^{k-1} + 1 \\ &= 2^k - 1. \end{aligned}$$

We can now apply the principle of mathematical induction which shows that the formula is valid for all $n$.

**16.** (a) For $n = 1$ the formula is valid.

(b) Suppose that the formula is valid for $n = k$, that is, suppose that

$$u_k = (2k - 1)^2$$

We shall prove that it is then valid for $n = k + 1$ as well. In fact,

$$u_{k+1} = u_k + 8k = (2k - 1)^2 + 8k$$
$$= 4k^2 - 4k + 1 + 8k$$
$$= 4k^2 + 4k + 1 = (2k + 1)^2.$$

Applying the principle of mathematical induction, we find that it is true for all natural numbers $n$.

**17.** $\Delta u_n = u_{n+1} - u_n = (n + 1)^2 - n^2 = 2n + 1.$

**18.** We shall show that for all $n$ the following formula is valid:

$$u_1 + \Delta u_1 + \Delta u_2 + \ldots + \Delta u_{n-1} = u_n.$$

In fact, recalling the definition of the sequence of differences, we obtain

$$u_1 + \Delta u_1 + \Delta u_2 + \ldots + \Delta u_{n-1}$$
$$= u_1 + (u_2 - u_1) + (u_3 - u_2) + \ldots + (u_n - u_{n-1}).$$

All of the terms of the latter sum except $u_n$ are canceled, and thus we get the required formula.

It is clear that a similar formula is valid for $v_n$ as well. Let us now find the value of $u_n - v_n$:

$$u_n - v_n = (u_1 + \Delta u_1 + \Delta u_2 + \ldots + \Delta u_{n-1}) -$$
$$- (v_1 + \Delta v_1 + \Delta v_2 + \ldots + \Delta v_{n-1})$$
$$= u_1 - v_1 + \Delta u_1 - \Delta v_1 + \Delta u_2 -$$
$$- \Delta v_2 + \ldots + \Delta u_{n-1} - \Delta v_{n-1}.$$

But from the conditions of the problem we know that

$$\Delta u_1 = \Delta v_1, \quad \Delta u_2 = \Delta v_2, \ldots, \quad \Delta u_{n-1} = \Delta v_{n-1},$$

and therefore

$$u_n - v_n = u_1 - v_1.$$

From this it follows that if $u_1 \neq v_1$, then $u_n - v_n \neq 0$, that is, that $u_n$ and $v_n$ are not equal for any $n$. If $u_1 = v_1$, then the equality $u_n = v_n$ is satisfied for all $n$.

**19.** The problem is solved by a direct verification:

$$\Delta w_n = w_{n+1} - w_n = (u_{n+1} + v_{n+1}) - (u_n + v_n)$$
$$= (u_{n+1} - u_n) + (v_{n+1} - v_n) = \Delta u_n + \Delta v_n.$$

**20.** We have that $\Delta u_n = (n + 1)^k - n^k$. We shall show that

$$(n + 1)^k = n^k + kn^{k-1} + \ldots,$$

where the dots signify terms containing $n$ raised to a power less than $k - 1$.

We shall prove this by induction on the number $k$.

(a) For $k = 1$ the assertion is valid.

(b) Suppose that the assertion is valid for $k = k_0$, that is, suppose that

$$(n + 1)^{k_0} = n^{k_0} + k_0 n^{k_0 - 1} + \ldots,$$

where the dots signify terms containing $n$ raised to a power less than $k_0 - 1$. We shall show that the assertion will then be valid for $k = k_0 + 1$ as well. We have

$$\begin{aligned}(n + 1)^{k_0+1} &= (n + 1)^{k_0}(n + 1)\\ &= (n^{k_0} + k_0 n^{k_0-1} + \ldots)(n + 1)\\ &= n^{k_0+1} + k_0 n^{k_0} + \ldots + n^{k_0} +\\ &\quad + k_0 n^{k_0-1} + \ldots\\ &= n^{k_0+1} + (k_0 + 1)n^{k_0} + \ldots,\end{aligned}$$

where the dots in the last expression signify terms containing $n$ raised to a power less than $k_0$. Thus our assertion is valid for $k = k_0 + 1$ as well.

Now, using the principle of mathematical induction, we obtain for all $k$,

$$(n + 1)^k = n^k + kn^{k-1} + \ldots.$$

Therefore,

$$\begin{aligned}\Delta u_n &= (n + 1)^k - n^k = n^k + kn^{k-1} + \ldots - n^k\\ &= kn^{k-1} + \ldots,\end{aligned}$$

where the dots signify terms containing $n$ raised to a power less than $k - 1$. This implies that $\Delta u_n$ is given by a polynomial of degree $k - 1$ in $n$ and that the coefficient of the leading term of this polynomial is equal to $k$.

**21.** Suppose that

$$u_n = a_0 n^k + a_1 n^{k-1} + \dots + a_{k-1} n + a_k.$$

We consider the sequence $u_n$ as the sum of $k + 1$ sequences with general terms $a_0 n^k, a_1 n^{k-1}, \dots, a_{k-1} n, a_k$, respectively. Then from the result of Problem 19, the sequence $\Delta u_n$ will be the sum of the sequence of differences of these $k + 1$ sequences. From Problem 20 it is clear that the general term of the first sequence of differences is given by a polynomial of degree $k - 1$ in $n$ with leading coefficient $a_0 k$, and that the general terms of the remaining sequences of differences will be given by polynomials of degree less than $k - 1$. Therefore the general term of the sequence of differences $\Delta u_n$ is given by a polynomial of degree $k - 1$ in $n$ with leading coefficient $a_0 k$.

**22.** We shall show that the $k$th sequence of differences of the sequence $u_n = n^k$ is equal to $k!$. We apply induction on $k$.

(a) It is clear that for $k = 1$ the assertion is valid, because if $u_n = n$, then $\Delta u_n = 1 = 1!$.

(b) Suppose that the assertion has been proved for all $k \leq k_0$. We shall prove it for $k = k_0 + 1$. Let us consider the sequence $u_n = n^{k_0 + 1}$. We know (see Problem 20) that the general term of its first sequence of differences is given by a polynomial of degree $k_0$ in $n$ with leading coefficient $k_0 + 1$. Suppose that this polynomial is

$$\Delta u_n = (k_0 + 1)n^{k_0} + a_1 n^{k_0 - 1} + a_2 n^{k_0 - 2} + \dots + a_{k_0}.$$

The sequence $\Delta u_n$ is the sum of the $k_0 + 1$ sequences with general terms $(k_0 + 1)n^{k_0}, a_1 n^{k_0 - 1}, \dots, a_{k_0}$, respectively. We are required to take the sequence of differences of the sequence $\Delta u_n$ $k_0$ times. Clearly, it suffices to take the sequence of differences of each summand $k_0$ times and then to add the result. From the validity of our assertion for $k = k_0$ it follows that the first summand contributes a sequence all of whose terms are identically equal to $(k_0 + 1)k_0!$ when the $k_0$ sequences of differences are taken. The validity of our assertion for $k = k_0 - 1$ implies that upon taking sequences of differences of the second summand $k_0 - 1$ times we get a sequence all of whose terms are identically equal to $a_1(k_0 - 1)!$, and upon taking the sequence of differences one more time we get a sequence all of whose terms are zero. Analogously, the $k_0$th sequence of differences of the remaining summands contributes nothing. Consequently, all of the terms of the $(k_0 + 1)$st sequence of differences are equal to $(k_0 + 1)!$.

In this manner we have proved that the $k$th sequence of differences of the sequence $n^k$ is equal to $k!$ for all $k$.

**23.** We solve the problem by induction on $k$.

(a) Suppose that $k = 0$, that is, that $v_n = 1$ for all $n$. The existence of the desired sequence is then clear, for we need merely take $u_n = n$. We shall denote this sequence by $\{u_n^{(0)}\}$.

(b) Suppose that the assertion is proved for all $k \leq k_0$. Then there will exist sequences $\{u_n^{(0)}\}$, $\{u_n^{(1)}\}$, . . . , $\{u_n^{(k_0)}\}$ such that $\Delta u_n^{(0)} = 1$, $\Delta u_n^{(1)} = n$, . . . , $\Delta u_n^{(k_0)} = n^{k_0}$. We shall show that there exists a sequence $\{u_n^{(k_0+1)}\}$ such that $\Delta u_n^{(k_0+1)}) = n^{k_0+1}$. We can find this sequence in the following way. First, we take the sequence with general term $n^{k_0+2}$. We know that the general term of its sequence of differences is given by a polynomial of degree $k_0 + 1$ with leading coefficient equal to $k_0 + 2$ (see Problem 21). Therefore the general term of the sequence of differences for the sequence $n^{k_0+2}/(k_0 + 2)$ is given by a polynomial in $n$ of degree $k_0 + 1$ having leading coefficient 1.

Suppose that this polynomial is

$$n^{k_0+1} + a_0 n^{k_0} + a_1 n^{k_0-1} + \dots + a_{k_0}.$$

Then as the sequence $\{u_n^{(k_0+1)}\}$ we take the sequence with the following general term:

$$u_n^{(k_0+1)} = \frac{n^{k_0+1}}{k_0 + 1} - a_0 u_n^{k_0} - a_1 u_n^{(k_0-1)} - \dots - a_{k_0} u_n^{(0)}.$$

We shall show that this sequence will satisfy the required conditions. In fact, using the results of Problem 19, we obtain

$$\Delta u_n^{(k_0+1)} = \Delta\left(\frac{n^{k_0+2}}{k_0+2}\right) - \Delta a_0 u_n^{k_0} - \Delta(a_1 u_n^{k_0-1}) - \\ - \ldots - \Delta(a_{k_0} u_n^0) \\ = n^{k_0+1} + a_0 n^{k_0} + a_1 n^{k_0-1} + \ldots + a_{k_0} - \\ - a_0 n^{k_0} - a_1 n^{k_0-1} - \ldots - a_{k_0} = n^{k_0+1}.$$

Thus there exists a sequence $\{u_n^{(k_0+1)}\}$ having the necessary properties. Using the principle of mathematical induction, we find that for every $k$ there exists a sequence $u_n^{(k)}$ for which $\Delta u_n^{(k)} = n^k$. It is also clear that the general term of the sequence $\{u_n^{(k)}\}$ is given by a polynomial in $n$ of degree $k + 1$ with leading coefficient $1/(k + 1)$.

**24.** We use the notation $v_n = \Delta u_n$, $w_n = \Delta v_n$, $x_n = \Delta w_n$ and $y_n = \Delta x_n$. We are given that $y_n = 0$ for all $n$. Therefore $x_n$ is a constant for all $n$. From this fact it is clear that for all $n$, $w_n$ is given by a polynomial in $n$ of degree 1. Suppose that this polynomial is $a_0 n + a_1$. From the solution of Problem 18 we know that

$$v_n = v_1 + w_1 + w_2 + \ldots + w_{n-1} \\ = v_1 + a_0[1 + 2 + \ldots + (n-1)] + a_1(n-1).$$

But we know that $1 + 2 + \ldots + (n-1) = \dfrac{n(n-1)}{2}$ (see page 4) and therefore that

$$v_n = v_1 + a_1(n-1) + a_0\frac{n(n-1)}{2};$$

that is, the general term of the sequence $v_n$ is given by a polynomial of degree 2 in $n$. Suppose that this polynomial is $v_n = b_0 n^2 + b_1 n + b_2$. Then

$$u_n = u_1 + v_1 + v_2 + \ldots + v_{n-1} \\ = u_1 + b_2(n-1) + b_1\frac{n(n-1)}{2} + \\ + b_0[1^2 + 2^2 + \ldots + (n-1)^2].$$

We shall show that the sum $1^2 + 2^2 + \ldots + (n-1)^2$ can be written in the form of a polynomial in $n$ of the third degree. In fact, if we denote this sum by $S_n$, then $\Delta S_n = n^2$. From Problem 23 it follows that $S_n$ is given by a polynomial in $n$ of the third degree. Thus, $u_n$ is also given by a polynomial in $n$ of the third degree.

**25.** Suppose that the polynomial $v_n$ has the form

$$v_n = a_0 n^k + a_1 n^{k-1} + \ldots + a_{k-1} n + a_k.$$

Let us take the sequence

$$w_n = a_0 u_n^{(k)} + a_1 u_n^{(k-1)} + \ldots + a_{k-1} u_n^{(1)} + a_k u_n^{(0)},$$

where $u_n^{(k)}, u_n^{(k-1)}, \ldots, u_n^{(1)}, u_n^{(0)}$ are the sequences whose existence was shown in Problem 23. From Problem 19 it follows that for all $n$, $\Delta w_n = v_n$. Thus $\Delta w_n = \Delta u_n$ for all $n$. From Problem 18a it follows that the differences $u_n - w_n$ are identical for all $n$ (they are all equal to $u_1 - w_1$). The equality

$$u_n = (a_0 u_n^{(k)} + a_1 u_n^{(k-1)} + \ldots + a_{k-1} u_n^{(1)} + a_k u_n^{(0)}) + \\ + (u_1 - w_1)$$

is therefore valid for all $n$. It is known from Problem 23 that $u_n^{(k)}$ is given by a polynomial in $n$ of degree $k + 1$. Therefore the general term of the sequence $u_n$ is given by a sequence of degree $k + 1$ in $n$.

**26.** Suppose that $u_n = 1 + 2^2 + \ldots + (n-1)^2 + n^2$. It is then clear that $\Delta u_n = n^2$. From Problem 24 it follows that $u_n$ is given by a polynomial of degree 3, that is, by

$$u_n = a_0 n^3 + a_1 n^2 + a_2 n + a_3.$$

We shall now find $a_0$, $a_1$, $a_2$, $a_3$. Substituting the numbers 0, 1, 2, and 3 for $n$, we get

$$\begin{aligned} u_0 &= a_3 = 0, \\ u_1 &= a_0 + a_1 + a_2 + a_3 = 1, \\ u_2 &= 8a_0 + 4a_1 + 2a_2 + a_3 = 5, \\ u_3 &= 27a_0 + 9a_1 + 3a_2 + a_3 = 14. \end{aligned}$$

We solve the system of equations that is obtained:

$$\left.\begin{aligned} a_0 + a_1 + a_2 &= 1, \\ 8a_0 + 4a_1 + 2a_2 &= 5, \\ 27a_0 + 9a_1 + 3a_2 &= 14. \end{aligned}\right\}$$

To do this we subtract three times the first equation from the third and two times the first from the second. We then get

$$\left.\begin{aligned} 24a_0 + 6a_1 &= 11, \\ 6a_0 + 2a_1 &= 3; \end{aligned}\right\}$$

$$6a_0 = 2, \quad a_0 = \tfrac{1}{3}, \quad a_1 = \tfrac{1}{2}, \quad a_2 = \tfrac{1}{6}.$$

Thus,

$$u_n = \tfrac{1}{3}n^3 + \tfrac{1}{2}n^2 + \tfrac{1}{6}n = \frac{n(2n^2 + 3n + 1)}{6}$$
$$= \frac{n(n + 1)(2n + 1)}{6};$$

that is,

$$1^2 + 2^2 + \ldots + n^2 = \frac{n(n + 1)(2n + 1)}{6}.$$

**Remark.** We did not need to use the results of Problem 24. Then, given the answer, it would have been necessary to prove it by the method of mathematical induction.

**27.** Let $u_n = 1 \cdot 2 + 2 \cdot 3 + \ldots + n(n + 1)$.

Then $\Delta u_n = n(n + 1)$ is a polynomial of the second degree. From the result in Problem 24 it follows therefore that $u_n$ is a polynomial of degree 3. Set

$$u_n = a_0 n^3 + a_1 n^2 + a_2 n + a_3.$$

To compute the numbers $a_0, a_1, a_2, a_3$ we substitute the values $n = 0, 1, 2$, and 3. We then get a system of four equations in four unknowns:

$$\left.\begin{aligned} u_0 &= a_3 = 0, \\ u_1 &= a_0 + a_1 + a_2 + a_3 = 2, \\ u_2 &= 8a_0 + 4a_1 + 2a_2 + a_3 = 8, \\ u_3 &= 27a_0 + 9a_1 + 3a_2 + a_3 = 20, \end{aligned}\right\}$$

which leads to the system

$$\left.\begin{aligned} a_0 + a + a &= 2, \\ 8a_0 + 4a_1 + 2a_2 &= 8, \\ 27a_0 + 9a_1 + 3a_2 &= 20. \end{aligned}\right\}$$

We subtract the first equation from the second and three times the first from the third. The result is

$$\left.\begin{aligned} 3a_0 + a_1 &= 2, \\ 24a_0 + 6a_1 &= 14. \end{aligned}\right\}$$

Subtracting six times the first equation from the second, we get

$$6a_0 = 2; \quad a_0 = \tfrac{1}{3}; \quad a_1 = 1; \quad a_2 = \tfrac{2}{3}.$$

Therefore,

$$u_n = \tfrac{1}{3}n^3 + n^2 + \tfrac{2}{3}n = \frac{n^3 + 3n^2 + 2n}{3} = \frac{n(n+1)(n+2)}{3};$$

that is,

$$1 \cdot 2 + 2 \cdot 3 + \ldots + n(n+1) = \frac{n(n+1)(n+2)}{3}.$$

**28.** (a) From the definition of an arithmetic progression we have

$$u_n = u_{n-1} + d = u_{n-2} + 2d = \ldots = u_1 + (n-1)d.$$

(b) From the definition of a geometric progression we have

$$u_n = u_{n-1}\, q = u_{n-2}\, q^2 = \ldots = u_1\, q^{n-1},$$

**29.** (a) We are required to find the following sum:

$$S_n = u_1 + u_2 + \ldots + u_{n-1} + u_n.$$

From Problem 28a we know that this sum can be written in the following way:

$$S_n = u_1 + (u_1 + d) + (u_1 + 2d) + \ldots + [u_1 + (n-2)d] + [u_1 + (n-1)d].$$

We can also write the sum with the terms in the reverse order:

$$S_n = [u_1 + (n-1)d] + [u_1 + (n-2)d] + \ldots + \\ + (u_1 + 2d) + (u_1 + d) + u_1.$$

Adding the two equalities obtained, we have

$$2S_n = \{u_1 + [u_1 + (n-1)d]\} + \\ + \{u_1 + d + [u_1 + (n-2)d]\} + \ldots + \\ + \{[u_1 + (n-2)d] + (u_1 + d)\} + \\ + \{[u_1 + (n-1)d] + u_1\}.$$

(Each term of the upper equality is at first added to the term of the second that is directly below it.) It is clear that the sum in each of the braces is equal to $2u_1 + (n-1)d$. There are $n$ of these braces. Therefore

$$2S_n = n[2u_1 + (n-1)d] = 2u_1 n + n(n-1)d, \text{ or}$$

$$S_n = u_1 n + \frac{n(n-1)d}{2}.$$

(b) We are required to find

$$P_n = u_1 \cdot u_2 \cdot \ldots \cdot u_{n-1} \cdot u_n.$$

From Problem 28b we know that $P_n$ can be written in the following form:

$$P_n = u_1 \cdot u_1 q \cdot u_1 q^2 \cdot \ldots \cdot u_1 q^{n-2} \cdot u_1 q^{n-1} \\ = u_1^n q^{1+2+\ldots+(n-2)+(n-1)}$$

But we know (see page 4) that $1 + 2 + \ldots + (n-2) + (n-1) = n(n-1)/2$. Therefore

$$P_n = u_1^n \cdot q^{\frac{n(n-1)}{2}}.$$

**30.** (a) According to the formula obtained in Problem 29a,

$$S_n = u_1 n + \frac{n(n-1)d}{2} = \frac{15 \cdot 14}{2} \cdot \frac{1}{3} = 35.$$

(b) According to the formula obtained in Problem 29b,

$$P_n = u_1^n q^{\frac{n(n-1)}{2}} = (\sqrt[3]{10})^{\frac{15\cdot 14}{2}} = 10^{\frac{15\cdot 14}{2}\cdot\frac{1}{3}} = 10^{35}.$$

**31.** (a) We know that for an arithmetic progression

$$u_1 = u_3 - 2d, \quad u_2 = u_3 - d, \quad u_4 = u_3 + d,$$
$$u_5 = u_3 + 2d.$$

Therefore the sum of the first five terms is

$$\begin{aligned} S_5 &= u_1 + u_2 + u_3 + u_4 + u_5 \\ &= (u_3 - 2d) + (u_3 - d) + u_3 + (u_3 + d) + (u_3 + 2d) \\ &= 5u_3 = 0. \end{aligned}$$

(b) We need to find

$$P_5 = u_1 \cdot u_2 \cdot u_3 \cdot u_4 \cdot u_5.$$

We know that for a geometric progression

$$u_1 = \frac{u_3}{q^2}, \quad u_2 = \frac{u_3}{q}, \quad u_4 = u_3 q, \quad u_5 = u_3 q^2.$$

Therefore,

$$P_5 = \frac{u_3}{q^2} \cdot \frac{u_3}{q} \cdot u_3 \cdot u_3 \cdot q \cdot u_3 q^2 = u_3^5 = 4^5 = 1024.$$

**32.** We are required to find the sum of the first $n$ terms of a geometric progression, that is, the quantity

$$\begin{aligned} S_n &= u_1 + u_2 + u_3 + \ldots + u_n \\ &= u_1 + u_1 q + u_1 q^2 + \ldots + u_1 q^{n-1}. \end{aligned}$$

We consider the quantity $qS_n - S_n$:

$$\begin{aligned} qS_n - S_n &= q(u_1 + u_1 q + u_1 q_2 + \ldots + u_1 q^{n-1}) \\ &\quad - (u_1 + u_1 q + u_1 q_2 + \ldots + u_1 q^{n-1}) \\ &= u_1 q + u_1 q^2 + u_1 q^3 + \ldots + u_1 q^{n-1} + u_1 q^n \\ &\quad - u_1 - u_1 q - u_1 q^2 - \ldots - u_1 q^{n-1}. \end{aligned}$$

We see that all of the terms in this sum cancel one another except for the first and the last; that is,

$$qS_n - S_n = u_1 q^n - u_1,$$
$$S_n(q - 1) = u_1 q^n - u_1,$$
$$S_n = \frac{u_1 q^n - u_1}{q - 1}.$$

**33.** It is easy to see that 2 more people will have been given the news within 10 minutes, 4 more after 20 minutes, and in general, $2^k$ more after $10k$ minutes.

The total number of people who will have heard the news by the end of $10k$ minutes will be equal to the number who will have heard it by the end of the first minute (that is, one person) plus the number who will have heard it between the end of the first minute and the end of the first 10 minutes (2 people) plus the number who will have heard it between the 10th minute and the 20th (4 people), and so on, up to the number of people who will have heard the news between the $10(k - 1)$st minute and the $10k$th minute ($2^k$ people).

Therefore this number will be equal to

$$S_k = 1 + 2 + 4 + \ldots + 2^k.$$

We see that $S_k$ is the sum of the first $k + 1$ terms of the geometric series with first term 1 and ratio 2. Therefore

$$S_k = \frac{1 \cdot 2^{k+1} - 1}{2 - 1} = 2^{k+1} - 1.$$

For the solution of the problem it remains only to find the smallest $k$ for which $S_k$, that is, the total number of people who will have heard the news at the end of the first $10k$ minutes, will be greater than 3,000,000 (the total number of inhabitants of the city). It is easy to see that $k = 21$ is the desired number, because

$$2^{22} - 1 = 4{,}194{,}303.$$

Thus all of the inhabitants of the city will have heard the news after 210 minutes or $3\frac{1}{2}$ hours.

**34.** We denote by $a_1, a_2, a_3, a_4$, and $a_5$ the amount of time which the cyclist spends on the first, second, third, fourth, and fifth laps, respectively, and by $b_1$, $b_2, b_3, b_4$, and $b_5$ the same quantities for the horseman. We know from the statement of the problem that the numbers $a_1, \ldots, a_5$ form a geometric progression with ratio 1.1 and that the numbers $b_1, \ldots, b_5$ form an arithmetic progression with increment $d$. We know that $a_1 = b_1$ and that $a_1 + a_2 + a_3 + a_4 + a_5 = b_1 + b_2 + b_3 + b_4 + b_5$. Using the formula that we obtained in Problems 29 and 32, we can rewrite the latter equation in the following way:

$$\frac{a_1(1.1)^5 - a_1}{1.1 - 1} = 5a_1 + \frac{5 \cdot 4d}{2},$$

$$a_1[10(1.1^5 - 1) - 5] = 10d, \quad \frac{d}{a_1} = \frac{10(1.1^5 - 1) - 5}{10}.$$

We need to find the ratio $b_5/a_5$:

$$\frac{b_5}{a_5} = \frac{a_1 + 4d}{a_1 \cdot 1.1^4} = \frac{1}{1.1^4} + \frac{4}{1.1^4} \cdot \frac{d}{a_1}.$$

Substituting the value of $d/a_1$ (which we know) in this equation, we obtain

$$\frac{b_5}{a_5} = \frac{1}{1.1^4} + \frac{4}{1.1^4} \cdot \frac{10(1.1^5 - 1) - 5}{10}$$

$$= \frac{1}{1.1^4}[1 + 4(1.1^5 - 1) - 2]$$

$$= \frac{1}{1.1^4}(4 \cdot 0.6105 - 1)$$

$$= \frac{1.4421}{1.4641} = 0.986 < 1.$$

Thus, we see that on the last lap the cyclist spends approximately 0.986 times less time than the horseman.

**35.** We are required to find the sum $1 + 3 + 5 + \ldots + 997 + 999$. Clearly, this is the sum of the first 500 terms of the arithmetic progression with first term 1 and increment 2. Therefore our sum is equal to

$$1\cdot 500 + \frac{500\cdot 499}{2}\cdot 2 = 500 + 500\cdot 499 = 500\cdot 500$$
$$= 250{,}000.$$

**36.** Using the sum $S^{(1)}$ of all positive three-digit numbers, we shall compute the sum $S^{(2)}$ of all positive numbers of three digits which are divisible by 2 and the sum $S^{(3)}$ of all positive numbers of three digits which are divisible by 3. In this computation the numbers divisible by both 2 and 3, that is, divisible by 6, will appear twice.

Therefore, to obtain our answer we shall have to add to the difference $S^{(1)} - S^{(2)} - S^{(3)}$ the sum $S^{(6)}$ of all positive numbers of three digits divisible by 6, that is, take the quantity

$$S = S^{(1)} - S^{(2)} - S^{(3)} + S^{(6)}.$$

Each of the four terms on the right in this equation is the sum of a certain number of terms of certain arithmetic progressions. The number of terms used and the increments of the progressions are shown in the following table.

| | $u_1$ | $n$ | $d$ |
|---|---|---|---|
| I | 100 | 900 | 1 |
| II | 100 | 450 | 2 |
| III | 102 | 300 | 3 |
| IV | 102 | 150 | 6 |

Therefore $S^{(1)} = 100\cdot 900 + \frac{900\cdot 899}{2}\cdot 1 = 90{,}000 + 404{,}550 = 494{,}550$,

$$S^{(2)} = 100\cdot 450 + \frac{450\cdot 449}{2}\cdot 2 = 247{,}050,$$

$$S^{(3)} = 102\cdot 300 + \frac{300\cdot 299}{2}\cdot 3 = 30{,}600 + 134{,}550$$
$$= 165{,}150,$$
$$S^{(6)} = 102\cdot 150 + 150\cdot 149\cdot 6 = 82{,}350,$$
$$S = S^{(1)} - S^{(2)} - S^{(3)} + S^{(6)} = 164{,}700.$$

**37.** The $n$th term of a sequence is obtained if the sum of the first $n - 1$ terms is subtracted from the sum of the first $n$ terms; that is,

$$u_n = S_n - S_{n-1}.$$

In our case,

$$u_n = S_n - S_{n-1} = 3n^2 - 3(n - 1)^2 = 6n - 3.$$

We now find $u_n - u_{n-1}$:

$$u_n - u_{n-1} = (6n - 3) - [6(n - 1) - 3] = 6.$$

Thus, the difference of any two consecutive terms of the sequence $u_n$ is 6. Therefore our sequence is an arithmetic progression with increment 6. The first term $u_1$ is equal to $S_1 = 3$.

**38.** Let us suppose that we have already found a geometric progression of the type in which we are interested. Suppose that the number 27 is in the $m$th place, 8 in the $n$th place, and 12 in the $p$th place. We shall denote the first term of the sequence by $u_1$ and its ratio by $q$. We then get

$$27 = u_1 q^{m-1},$$
$$8 = u_1 q^{n-1},$$
$$12 = u_1 q^{p-1}.$$

Dividing the first equality by the second and then the third, we get

$$\tfrac{27}{8} = q^{m-n},$$
$$\tfrac{9}{4} = q^{m-n}.$$

Raising the first equation to the $(m - p)$th power and the second to the $(n - p)$th and equating their right-hand sides, we get

$$(\tfrac{27}{8})^{m-p} = (\tfrac{9}{4})^{m-n},$$
$$3^{3m-3p-2m+2n} = 2^{3m-3p-2m+2n},$$
$$3^{m-3p+2n} = 2^{m-3p+2n}.$$

It is not difficult to see that the last inequality can be satisfied only if $m - 3p + 2n = 0$. Conversely, suppose that $m$, $n$, and $p$ are positive integers such that $m - 3p + 2n = 0$. Then the equalities

$$3^{m-3p+2n} = 2^{m-3p+2n},$$
$$(\tfrac{27}{8})^{m-p} = (\tfrac{9}{4})^{m-n}$$

are satisfied.

Suppose that $q = \sqrt[m-n]{\frac{27}{8}}$. Then

$$\tfrac{27}{8} = q^{m-n} \text{ and } \tfrac{9}{4} = q^{m-p}.$$

We denote by $u_1$ the number for which $12 = u_1 q^{p-1}$ (such a number always exists). Then from the equations $\frac{27}{8} = q^{m-n}$ and $\frac{9}{4} = q^{m-n}$ it follows that

$$8 = u_1 q^{n-1},$$
$$27 = u_1 q^{m-1}.$$

This means that in a geometric progression with ratio $q$ and first term $u_1$ the numbers 27, 8, and 12 are in the $m$th, $n$th, and $p$th places, respectively. As an example of such a sequence we take the sequence with $u_1 = 8$, $q = \frac{3}{2}$. Then $m = 4$, $n = 1$, $p = 2$.

**39.** We suppose that we have found a geometric progression in which the numbers 1, 2, and 5 lie in the $m$th, $n$th and $p$th positions, respectively. Then

$$1 = u_1 q^{m-1},$$
$$2 = u_1 q^{n-1},$$
$$5 = u_1 q^{p-1}.$$

Dividing the second and third equations by the first, we get

$$2 = q^{n-m},$$
$$5 = q^{p-m}.$$

Raising the first equation to the $(p - m)$th power and the second to the $(n - m)$th and equating the right sides, we obtain

$$2^{p-m} = 5^{n-m}.$$

It is clear that this equation can be satisfied only if $p - m = n - m = 0$, since otherwise the right side will be odd and the left even.

But from the statement of the problem it is clear that the numbers $m$, $n$, and $p$ must all be different. Therefore our assumption of the existence of a geometric progression with the required properties leads to a contradiction; and so such a progression does not exist.

**40.** We denote by $u_{12}$, $u_{13}$, and $u_{15}$ the 12th, 13th, and 15th terms of the arithmetic progression. We know that the numbers $u_{12}^2$, $u_{13}^2$, and $u_{15}^2$ form a geometric progression, that is, that

$$\frac{u_{13}^2}{u_{12}^2} = \frac{u_{15}^2}{u_{13}^2}.$$

Taking the square root of both sides of this equation, we find that either

$$\frac{u_{13}}{u_{12}} = \frac{u_{15}}{u_{13}},$$

or

$$\frac{u_{13}}{u_{12}} = -\frac{u_{15}}{u_{13}}.$$

Let us examine the first equality. For any arithmetic progression $u_{13} = u_{12} + d$, $u_{15} = u_{13} + 2d = u_{12} + 3d$, where $d$ is the increment of the progression. Therefore

$$\frac{u_{12} + d}{u_{12}} = \frac{u_{13} + 2d}{u_{13}},$$

$$1 + \frac{d}{u_{12}} = 1 + \frac{2d}{u_{13}},$$

$$\frac{d}{u_{12}} = \frac{2d}{u_{13}}.$$

Let us suppose that $d \neq 0$. Then

$$\frac{u_{13}}{u_{12}} = 2 \quad \frac{u_{13}^2}{u_{12}^2} = 4;$$

that is, in this case the ratio of the geometric progression that is obtained is equal to 4.

Analogously, we can consider the case where the equality

$$\frac{u_{13}}{u_{12}} = -\frac{u_{15}}{u_{13}}$$

is satisfied.

In this case, carrying out the same operations, we obtain:

$$u_{12}^2 + 4u_{12}d + d^2 = 0,$$

$$\left(\frac{d}{u_{12}}\right)^2 + 4\frac{d}{u_{12}} + 1 = 0,$$

$$\frac{d}{u_{12}} = -2 \pm \sqrt{3}.$$

Case a:

$$\frac{d}{u_{12}} = -2 + \sqrt{3}.$$

$$\frac{u_{13}}{u_{12}} = 1 + \frac{d}{u_{12}} = -1 + \sqrt{3},$$

$$q = \left(\frac{u_{13}}{u_{12}}\right)^2 = 4 - 2\sqrt{3}.$$

Case b:

$$\frac{d}{u_{12}} = -2 - \sqrt{3}.$$

$$\frac{u_{13}}{u_{12}} = 1 + \frac{d}{u_{13}} = -1 - \sqrt{3},$$

$$q = \left(\frac{u_{13}}{u_{12}}\right)^2 = 4 + 2\sqrt{3}.$$

Thus, from the equality

$$\frac{u_{13}}{u_{12}} = -\frac{u_{15}}{u_{13}}$$

it follows that $q$ can be equal either to $4 - 2\sqrt{3}$ or to $4 + 2\sqrt{3}$.

Only the case where $d = 0$ remains. It is clear that in this case $q = 1$.

**41.** Suppose that we have such a geometric progression. Then for this progression the equality

$$u_1 q^{n+1} = u_1 q^n + u_1 q^{n-1}$$

is satisfied for all $n$. From this it follows that

$$q^2 = q + 1,$$

$$q = \frac{1 \pm \sqrt{5}}{2}.$$

Thus, the ratio of a geometric progression for which each term is equal to the sum of the two preceding ones can have only the values $(1 + \sqrt{5})/2$ and $(1 - \sqrt{5})/2$. It is clear that the converse is also true, that is, that for any geometric progression with one of these numbers as ratio, each term is equal to the sum of the two preceding ones.

**42.** We denote the first term of the first progression by $u_1$, its ratio by $q$, the first term of the second progression by $v_1$, and its ratio by $p$. We know that

$$u_1 + v_1 = 0,$$
$$u_1 q + v_1 p = 0.$$

We are to find $u_1 q^2 + v_1 p^2$:

$$\begin{aligned} u_1 q^2 + v_1 p^2 &= u_1 q^2 - u_1 p^2 = u_1(q^2 - p^2) \\ &= u_1(q - p)(q + p). \end{aligned}$$

But from the equalities

$$u_1 + v_1 = 0,$$
$$u_1 q + v_1 p = 0,$$

it follows that $u_1(q - p) = 0$.

Therefore, $u_1q^2 + v_1p^2 = u_1(q - p)(q + p) = 0$; that is, the third term of our sequence is also equal to 0.

**43.** We are to find two geometric progressions $\{u_n\}$ and $\{v_n\}$ for which the following equalities are satisfied:

$$u_1 + v_1 = 1,$$
$$u_2 + v_2 = 1,$$
$$u_n + v_n = u_{n-1} + v_{n-1} + u_{n-2} + v_{n-2}.$$

We denote the ratios of these progressions by $p$ and $q$. It is clear that $p \neq q$, since the sum of two geometric progressions with the same ratios is a geometric progression, but the Fibonacci sequence is not a geometric progression. Our last equality can be rewritten as follows:

$$u_1q^{n-1} + v_1p^{n-1} = u_1q^{n-2} + v_1p^{n-2} + u_1q^{n-3} + v_1p^{n-3},$$
$$u_1q^{n-3}(q^2 - q - 1) = -v_1p^{n-3}(p^2 - p - 1).$$

We shall show that $p^2 - p - 1 = 0$ and $q^2 - q - 1 = 0$. In fact, suppose that this is not so and suppose, for example, that $p^2 - p - 1 \neq 0$. Then, dividing both sides of the last equality by

$$v_1q^{n-3}(p^2 - p - 1),$$

we obtain

$$\frac{u_1}{v_1} \cdot \frac{q^2 - q - 1}{p^2 - p - 1} = -\left(\frac{p}{q}\right)^{n-3} \quad \text{for all } n \geq 3.$$

The left side of this equality is independent of $n$. Therefore, the right side must also be independent of $n$. But this can happen only if $p = q$. We have, however, already seen that $p \neq q$. Thus, we have arrived at a contradiction, that is, our assumption that $p^2 - p - 1 \neq 0$ is false. Therefore, $p^2 - p - 1 = 0$. Analogously, $q^2 - q - 1 = 0$. From this it follows that $p$ and $q$ are the roots of the equation $x^2 - x - 1 = 0$, which are distinct. The roots of this equation are $(1 + \sqrt{5})/2$ and $(1 - \sqrt{5})/2$. Let us suppose, for example, that $p = (1 + \sqrt{5})/2$ and $q = (1 - \sqrt{5})/2$. We shall now find $u_1$ and $v_1$. We use the following two equations:

$$u_1 + v_1 = 1,$$

$$u_1 \frac{1 + \sqrt{5}}{2} + v_1 \frac{1 - \sqrt{5}}{2} = 1.$$

From these equations, we obtain

$$u_1 = \frac{1 + \sqrt{5}}{2\sqrt{5}}; \quad v_1 = \frac{-1 + \sqrt{5}}{2\sqrt{5}}.$$

Therefore

$$u_n = u_1 p^{n-1} = \frac{1 + \sqrt{5}}{2\sqrt{5}} \left(\frac{1 + \sqrt{5}}{2}\right)^{n-1}$$

$$= \frac{1}{\sqrt{5}} \left(\frac{1 + \sqrt{5}}{2}\right)^{n};$$

$$v_n = v_1 q^{n-1} = \frac{-1 + \sqrt{5}}{2\sqrt{5}} \left(\frac{1 - \sqrt{5}}{2}\right)^{n-1}$$

$$= -\frac{1}{\sqrt{5}} \left(\frac{1 - \sqrt{5}}{2}\right)^{n}.$$

Consequently,

$$a_n = u_n + v_n = \frac{1}{\sqrt{5}} \left[\left(\frac{1 + \sqrt{5}}{2}\right)^{n} - \left(\frac{1 - \sqrt{5}}{2}\right)^{n}\right].$$

**44. First Solution.** The number of ways of lighting the kitchen for which one of the five lamps is lit is equal to the number of ways of lighting the kitchen for which four of the five lamps are lit (Fig. 27).

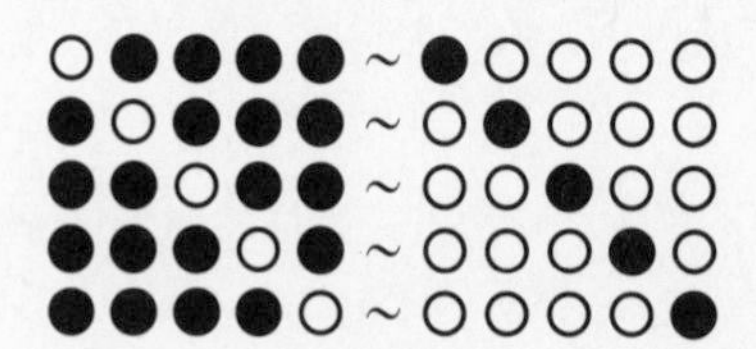

Fig. 27

Analogously, the number of ways of lighting the kitchen for which two of the five lamps are lit is equal to the number of ways for which two of the five lamps are not lit (that is, for which three of the five lamps are lit). The possibilities are shown in Fig. 28 (there are

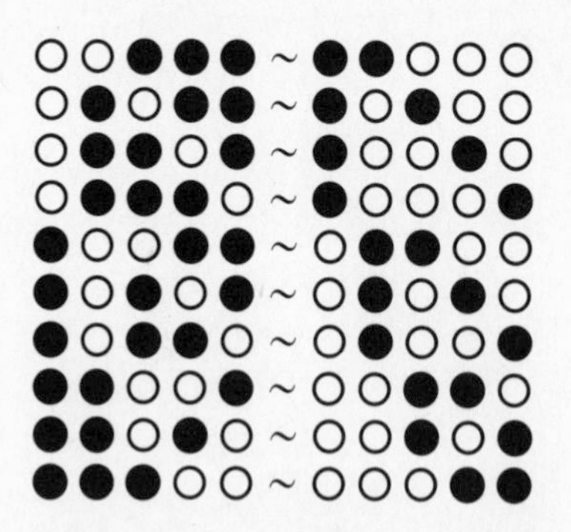

Fig. 28

All of the lamps are lit

○○○○○

None of the lamps are lit

Fig. 29

ten possibilities if two of the lamps are lit and ten possibilities if three of the lamps are lit). The two remaining ways of lighting the kitchen are shown in Fig. 29 (all of the lamps are lit; none of the lamps are lit). Having sorted out all of the possibilities, we can now count them easily.

**Second Solution.** (1) Suppose that there is only one lamp in the kitchen. This lamp can be in either of two states: lit ○ or not lit ● ; thus, there are two possibilities.

(2) Suppose we have 2 lamps. The first lamp can be in either of the two states ● or ○.

Each of these states can occur in combination with either of the two states of the second lamp: if the second lamp is not lit, we get two ways of lighting the kitchen.

If the second lamp is lit, we get two more ways of lighting the kitchen.

Thus there are four ways in all of lighting the kitchen:

(3) Suppose we have three lamps. The first two lamps can be in any one of four states. Each of these four states can occur in combination with either of the two states of the third lamp.

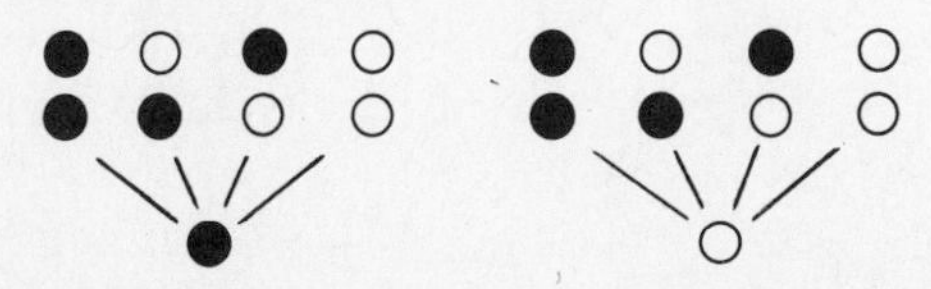

Thus there are eight ways in all of lighting the kitchen:

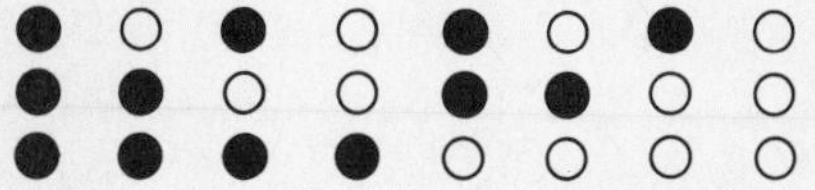

(4) Suppose we have four lamps. The first three lamps can occur in any one of eight states. Each of these states can occur in combination with either of the two states of the fourth lamp. Thus we have $8 \cdot 2 = 16$ ways of lighting the kitchen.

(5) Suppose we have five lamps. We have $16 \cdot 2 = 32$ ways of lighting the kitchen. The method of solution we are considering allows insignificant modifications: the step from two lamps to three can be illustrated not only by the scheme just shown, but by the following schemes as well:

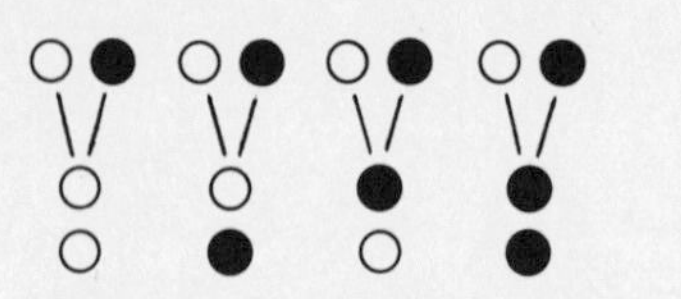

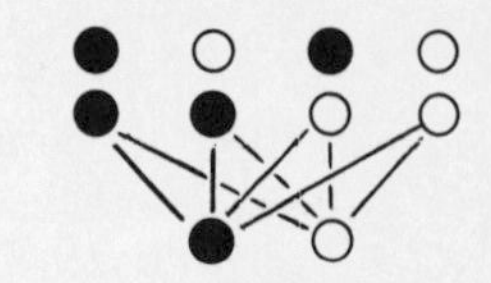

Both show that the addition of another lamp (which can be in either of two states) doubles the number of ways of lighting the kitchen.

**45.** How many ways of lighting a room are there if there are $n$ lamps in the room? We shall calculate the number of ways using two methods (compare the solution of Problem 44).

**First Method:**

0. Suppose not a single lamp is lit. There are $C_n^0$ such ways of illuminating the room (0 lamps are lit). It is clear that $C_n^0 = 1$ for any $n$. Still in order that the formula to be proved appear "elegant," we will write not 1 but $C_n^0$.

1. Exactly 1 of the $n$ lamps can be lit. The number of ways in which this can be done is $C_n^1$.

2. By $C_n^2$ we denote the number of ways of lighting the room for which two of the $n$ lamps are lit.

(k) When $k$ of the $n$ lamps are lit, we get $C_n^k$ ways of lighting the room.

(n) Finally, all $n$ of the lamps can be lit. The number of such ways of lighting the room is denoted by $C_n^n$. (It is clear that $C_n^n = 1$.)

Thus, the total number of ways of lighting the room is

$$C_n^0 + C_n^1 + C_n^2 + \ldots + C_n^k + \ldots + C_n^n.$$

**Second Method.** We denote the number of ways of lighting $n$ lamps (each of which can be either lit or not lit) by $D_n$. We prove by induction that $D_n = 2^n$ (in essence, this was already proved in the hint for this problem).

The number of ways of lighting a room with one lamp is $D_1$. Clearly, $D_1$ is equal to 2. Furthermore, $D_n = 2D_{n-1}$ (each way of lighting the $(n-1)$st lamp gives rise to 2 ways of lighting $n$ lamps: the $n$th lamp can be either lit or not lit). This means that from the assumption that $D_{n-1} = 2^{n-1}$ it follows that $D_n = 2^n$. Thus $D_n = 2^n$. But we have already shown that

$$D_n = C_n^0 + C_n^2 + \ldots + C_n^k + \ldots + C_n^n.$$

Consequently,

$$C_n^0 + C_n^1 + \ldots + C_n^k + \ldots + C_n^n = 2^n.$$

**46.** If there is only one traffic light, there are three ways of turning it on. Suppose that we now add a second traffic light. For each way of turning on one of the traffic lights, we can get three ways of turning on the two of them by changing the second light. This means that the number of possibilities has been tripled. Thus, there are $3 \cdot 3 = 3^2$ possibilities. Suppose that we now add a third traffic light. For each way of turning on two of the traffic lights we can again get three ways of turning on the three lights by changing the state of the third light. Again the number of possibilities is tripled. For three traffic lights there will be $3 \cdot 3^2 = 3^3$ possibilities. Upon the addition of an extra traffic light the number of possibilities will thus always be tripled. Therefore $n$ traffic lights can be lit in $3^n$ ways.

**47.** Suppose that the first $k$ traffic lights are lit in some fashion (there are $3^k$ such possibilities in all, see Problem 46). Thus, the remaining ones can be lit in

$2^{n-k}$ ways (see Problem 45). Then from each way of turning on the first $k$ traffic lights we shall be able to get $2^{n-k}$ ways of turning on all of the traffic lights. Therefore, there will be $3^k \cdot 2^{n-k}$ ways in all.

**49.** Answer: $10^4 \cdot 26^3$ numbers (in an alphabet of 26 letters).

**50.** Let us assign to each inhabitant of the kingdom a sequence of 32 zeros and ones in accordance with the following rule: If the inhabitant has his first tooth, then we place a 1 in the first place of the sequence, and if not, a 0; if the inhabitant has his second tooth, then we insert a 1 in the second place of the sequence, and if not, a 0; and so on (we suppose that the teeth have been numbered once and for all). From the conditions it is clear that each inhabitant will have been assigned a distinct sequence, and therefore the maximal number of inhabitants in the kingdom is simply equal to the number of such sequences. But the number of such sequences is $2^{32}$ (see the solution to Problems 44 and 45). Incidentally, $2^{32} \approx 4{,}000{,}000{,}000$ (four billion). This is more than the present population of the entire world.

**51.** We shall examine the terms that are obtained when the expression is expanded but before like terms are reduced. Each such term is the product of 100 factors, each of which is either the letter $x$ or a number. The first factor is either $x$ or $-1$, the second factor is either $x$ or $-2$, and so on. We are interested only in those products for which $x$ occurs 99 times and a number occurs only once. There will be a hundred of these, because a number can occur in the first place, in the second place, and so on, up to the hundredth place.

The corresponding products will be:

$$-1 \cdot \underbrace{x \cdot x \cdot \ldots \cdot x}_{99 \text{ times}}, \qquad x \cdot (-2) \cdot \underbrace{x \cdot x \cdot \ldots \cdot x}_{98 \text{ times}}, \ldots,$$

$$\underbrace{x \cdot x \cdot \ldots \cdot x \cdot}_{99 \text{ times}}(-100).$$

Thus the coefficient of $x^{99}$ will be equal to

$$-(1 + 2 + 3 + \ldots + 100) = -\frac{101 \cdot 100}{2} = -5050$$

(the parentheses contain the sum of the terms of an arithmetic progression. See Chapter 1).

**52. A. The order is essential.** In this case each representation is defined by the first summand, which can be any natural number less than $n$ (the second summand will then also be a natural number). Thus, there will be $n - 1$ representations in all.

**B. The order is inessential.** If $n$ is odd, then there will be one-half as many representations as in case A, since the representations $n = 1 + (n - 1)$ and $n = (n - 1) + 1, n = 2 + (n - 2)$ and $n = (n - 2) + 2$ and so on are now considered identical. The answer is thus $(n - 1)/2$. If $n$ is even, then commuting the summands in the representation $n = n/2 + n/2$ does not change its appearance, but commuting the summands in every other representation does. Thus we get the following answer:

$$1 + \frac{(n - 1) - 1}{2} = \frac{n}{2}.$$

**53.** We shall count how many representations there are for which the first summand is 1. The sum of the second and third summands must be $n - 1$. It is clear, then, that there are as many such representations as there are representations of the number $n - 1$ as the sum of two terms. But we already know that there are $n - 2$ such representations (see the solution to Problem 52). In the same way, the number of representations for which the first summand is 2 is equal to $n - 3$, and so on. There is one representation for which the first summand is $n - 2$. We can now count all of the representations. There will be

$$(n - 2) + (n - 3) + \ldots + 2 + 1 = \frac{(n - 2)(n - 1)}{2}.$$

(The left side is the sum of the terms of an arithmetic progression.) In the same way calculate the number of representations of the number $n$ as the sum of four terms. The answer to this problem is

$$\frac{(n-1)(n-2)(n-3)}{6}.$$

For another solution of Problem 53 read the remark to the solution of Problem 60.

**54.** We apply the method of mathematical induction. For $n = 1$ (the second line) the sum of the numbers is equal to 2. We shall prove that the sum of the numbers in the $(n + 1)$st line is twice that of the numbers in the $n$th line. To do this, we write down the $n$th and $(n + 1)$st lines of Pascal's triangle in the following form:

$$\begin{array}{ccccccc} 1 & a & b\ldots f & & g & 1 & \\ 0+1 & 1+a & a+b & \ldots & f+g & g+1 & 1+0. \end{array}$$

We calculate the sum of the numbers in the $(n + 1)$st line:

$$(0 + 1) + (1 + a) + (a + b) + \ldots +$$
$$+ (f + g) + (g + 1) + (1 + 0).$$

We add the first summands in each set of parentheses and then do the same for the second summands. We get

$$(0 + 1 + a + b + \ldots + f + g + 1) +$$
$$+ (1 + a + b + \ldots + g + 1 + 0)$$
$$= 2(1 + a + b + \ldots + f + g + 1).$$

The left side is the sum of the numbers in the $(n + 1)$st line; and the right side is twice the sum of the numbers in the $n$th line. It is now clear that the sum of the numbers in the $(n + 1)$st line is equal to $2^n$.

**55.** In the hint to the problem a concrete example of the arrangement of 14 bishops is shown. Consequently, 14 bishops can be placed on the board. If we prove that it is impossible to set up more than 14 bishops, then the first half of the problem will have been solved. We shall first consider bishops on black squares. How many can be placed on the chessboard so that they do not threaten one another? It is possible to set up 7 of them; we shall show that it is impossible to set up more. Let us take 7 black diagonals parallel to one another (Fig. 30). On each diagonal we can place at most one bishop. Therefore the answer is 7. We get the same answer if we take the 8 black diagonals perpendicular to those shown (the outer diagonals contain one square each). An examination of these diagonals convinces us that more than 7 bishops cannot be placed on the board (indeed, if 8 bishops are placed on the board so that they do not threaten one another, 2 of them must be on the outer diagonals; but as each of these diagonals consists of a single square, these 2 bishops will have to be located on opposite corners of the board and thus will threaten each other). But this argument is superfluous, as we get our result more simply if we choose our diagonals as shown in Fig. 30.

Fig. 30

The maximal number of bishops on white squares not threatening one another is also equal to 7. Bishops on white squares and bishops on black squares cannot threaten one another. Thus we can place at most 14 bishops on a chessboard in such a way that they do not threaten one another.

Let us denote by $W$ the number of ways in which we can place 7 bishops on white squares so that they do not threaten one another; by $B$, the number of ways of placing 7 bishops on black squares so that they do not threaten one another; and finally, by $T$, the number of ways of placing 14 bishops on the board so that they do not threaten one another. It is clear that $W = B$ and that $T = W \cdot B = (B)^2$, that is, $T$ is a square.

**Remark.** Solve a more general problem. Suppose that the board has dimensions $n \times n$, where $n$ is even. Prove that the greatest number of bishops that can be placed on the board without their threatening one another is equal to $(2n - 2)$. Prove that the number of ways of placing $2n - 2$ bishops on the board without their threatening one another is equal to $2^n$.

**56.** Here is one way of giving out the apples and pears: ○ ● ○ ● ● (on the first and third days an apple is given away, and on the second, fourth, and fifth days, a pear), and here is one more way: ● ● ● ○ ○ (on the first three days pears are given, and on the last two, apples). Thus, we must count all of the permutations of two light and three dark circles; but we have already done this in the solution to Problem 44. The answer is $C_5^2 = 10$.

**57.** For each way of distributing the fruit we make a table consisting of $k$ light and $n$ dark circles; the days on which apples are given away are denoted by the light circles, and the days on which pears are given away are denoted by the dark circles.

Thus, the problem is equivalent to the following:

Supposing that there are $n + k$ lamps, find the number of ways in which $k$ of them can be lit.

We need only recall that we have denoted this

number by $C^k{}_{n+k}$. (An explicit expression involving $n$ and $k$ can be obtained by comparing this answer with the results of Problems 60, 66, and 67.)

**59.** We have already proved that exactly one rook must lie on each column (see the hints). Exactly one rook will be on each row too, of course. One could place a rook on any one of the 8 squares of column $a$ (assuming that all of the rows are free). But after a rook is placed on the board, one row will be occupied (if, for example, the rook is on square $a2$, no other rook can be placed on row 2). Therefore there are only seven ways of setting up a rook on column $b$ (in the example just mentioned, the permissible squares are $b1, b3, b4, \ldots, b8$). Once the $b$ rook is in position, 6 free rows will remain (2 being occupied), so that there will be 6 squares on which the $c$ rook can be placed. Continuing in this way, we see that there will be 5 ways of setting up rook $d$, 4 ways for rook $l$, 3 ways for $f$, 2 for $g$, and only one for $h$. Since the number of ways of setting up rook $a$ was equal to 8, the number of ways of setting up rooks $a$ and $b$ is equal to $8 \cdot 7$; the number of ways of setting up rooks $a$, $b$, and $c$ is equal to $8 \cdot 7 \cdot 6$; and so on. Thus we can set up rooks $a, b, c, \ldots, h$ in $8 \cdot 7 \cdot 6 \cdot 5 \cdot 4 \cdot 3 \cdot 2 \cdot 1 = 8!$ ways (! denotes factorial; see Problem 5). To check whether you understand the solution solve this problem: "How many ways are there of setting up $n$ rooks on a board of $n \times n$ squares so that they do not threaten one another? The answer is $n!$. Here is another test question: a second solution of Problem 59 in which another answer is obtained (?!). If you believe that the first solution is true, find the error in the second solution.

The second solution will be given directly for an $n \times n$ board. We denote the number of ways of placing $n$ rooks on the board in the required manner by $L_n$. We shall prove that $L_n = (n!)^2$.

We carry out the proof by induction. For $n = 1$ the theorem is obviously true: $L_1 = (1!)^2$ (on a board of one square we can set up a rook in only one way). We shall show that if $L_{n-1} = [(n - 1)!]^2$, then

Rook stands on square P

Fig. 31

$L_n = (n!)^2$. For this it suffices to prove that $L_n = n^2 \cdot L_{n-1}$. Consider a single rook. We can set it up on the board on any one of $n^2$ squares. The remainder of the argument is illustrated by the figures, in which $n = 4$. Suppose that the first rook has been placed on one of the squares (square $P$ in Fig. 31). We remove from the board the row and column passing through this square (they are hatched in the figure). The remaining rooks cannot be placed on this row and column. Using the parts into which the board is now decomposed (denoted by $A$, $B$, $C$, and $D$ in the figure), we assemble a new board of dimensions $(n - 1) \times (n - 1)$ by shifting the pieces parallel to one another. On the original board we set up $n$ rooks so that they do not threaten one another, and so that the first rook sits on square $P$. None of the rooks other than the first will sit on any of the hatched squares. We join together parts $A$, $B$, $C$, and $D$ with the rooks in their places. On the board that is obtained, an arrangement of rooks will result for which the rooks do not threaten one another. To each arrangement of the rooks on the first board there corresponds a distinct arrangement of the rooks on the second board, and each arrangement of the rooks on the second board can be obtained in this way. There are $L_{n-1}$ arrangements in all on the second board; and the square $P$ can be chosen in $n^2$ ways. Thus, $L_n = n^2 \cdot L_{n-1}$.

Where is the error?

It is, perhaps, difficult to find the error. The proof, moreover, is not very simple. To make things easier, we shall give one more incorrect proof (for the case $n = 4$).

That is, we want to know how many ways there are of arranging 4 rooks on a $4 \times 4$ board so that the rooks do not threaten one another.

We can place the first rook on any of 16 squares. Once a rook has been placed on a square (see Fig. 31), we can place another rook only on the unhatched squares. Thus, there are nine ways in which we can place a second rook on the board. With the first two rooks in place, there will be only 4 squares remaining on which we can place another rook. With three rooks in place, there will be only one way of setting up the fourth rook (so that the rooks do not threaten one another). Thus, $L_4 = 16 \cdot 9 \cdot 4 \cdot 1 = (4!)^2$.

Fig. 32

Where is the error? Think about this before you read further.

This solution would be correct if the rooks had been numbered. Then the arrangements shown in Fig. 32 would be different. Since the rooks have not been numbered, however, these arrangements are identical (Fig. 33).

Fig. 33

Thus, we have proved the following two theorems:

1. The number $l_n$ of arrangements of $n$ rooks[1] on an $n \times n$ board is equal to $n!$ if the rooks are indistinguishable.
2. The number $L_n$ of arrangements of $n$ rooks[1] on an $n \times n$ board is equal to $(n!)^2$ if the rooks are numbered.

Each of these theorems implies the other. We again verify this only for the case $n = 4$ (generalize the argument to the case of arbitrary $n$ as an exercise).

We mark off 4 squares on a $4 \times 4$ board satisfying the condition that if 4 rooks are placed on these squares, they do not threaten one another. We call these squares $a$, $b$, $c$, and $d$. It is clear that the number of different ways of choosing 4 such squares is equal to $l_4$. We write numbers 1, 2, 3, and 4 on the rooks. Let $P_4$ be the number of distinct arrangements of rooks 1, 2, 3, and 4 on squares $a$, $b$, $c$, and $d$. The number $P_4$, of course, is independent of the choice of the squares: If $A$, $B$, $C$, and $D$ are any other squares, the number of distinct arrangements of rooks 1, 2, 3, and 4 on squares $A$, $B$, $C$, and $D$ is also equal to $P_4$. Therefore, $L_4 = P_4 \cdot l_4$. All that remains is to compute the number $P_4$. This you can do yourself. The answer is $P_4 = 4!$. Thus, $L_4/l_4 = 4!$. The general formula $L_n/l_n = n!$ is also true (verify this). Therefore, knowing that $l_n = n!$, we

[1]Not threatening one another.

find that $L_n = (n!)^2$; and conversely, knowing that $L_n = (n!)^2$, we find that $l_n = n!$. We have used before the symbol $P_n$; $P_n$ is the number of ways of permuting $n$ numbered rooks on $n$ fixed squares of the board.

**60.** We enumerate the $n$ objects as follows: $1, 2, \ldots, n$. Each way of choosing 2 out of $n$ objects now corresponds to a pair of numbers (the numbers of the objects chosen); the pairs $(k, m)$ and $(m, k)$, clearly, define the same choice. Thus, $C_n^2$ is equal to the number of pairs $(1, 2), (1, 3), \ldots, (n-1, n)$.

The first number in each pair can assume $n$ values; the second can assume only the remaining $n - 1$ values. Thus there are $n(n - 1)$ pairs in all. But in this calculation the pairs of numbers $(k, m)$ and $(m, k)$ have both been counted even though they define the same pair of objects. Thus,

$$C_n^2 = \frac{n(n-1)}{2}.$$

**Remark.** We can now give a new solution of Problem 53.

The decomposition of the number $n$ as the sum of 3 terms can be viewed as follows: We can suppose that we have a spoke (as in an abacus) on which $n$ beads are fastened, and we wish to divide the beads in two places. There are $n - 1$ intervals between the beads. There are $C_{n-1}^2$ ways of choosing two of these intervals and dividing the beads there. Thus the answer is

$$C_{n-1}^2 = \frac{(n-1)(n-2)}{2}.$$

**62.** We denote by $P_8^5$ the number of ways of distributing 5 pairwise distinct apples to the eight sons. We compute $P_8^5$ in two ways.

**First Method.** The number of ways of choosing five sons (those who are to get the oranges) is equal to $C_8^5$. To each such choice there correspond $5!$ ways of distributing the oranges (the number of ways of distributing 5 pairwise distinct oranges to five sons). Thus, $P_8^5 = C_8^5 \cdot 5!$.

**Second Method.** The first orange can be given to any one of the eight sons; the second, to any one of the remaining seven sons, and so on. The fifth can be given to any of the four sons who remain (the remaining three will not get apples). Thus, $P_8^5 = 8\cdot 7\cdot 6\cdot 5\cdot 4$ $8!/3!$.

**63.** We shall prove that $C_n^k$ is the number in the $(k + 1)$st place of the $(n + 1)$st line of Pascal's triangle. We shall prove this by induction on $n$. For $n = 1$ the theorem is true; for the entry in the first place of the second line of Pascal's triangle is $1 = C_1^0$, and the entry in the second place is $1 = C_1^1$. Suppose that the theorem is true for $n$. We shall show that it is then also true for $n + 1$. If $k = 0$ or $k = n + 1$, the theorem is true since $C_{n+1}^0 = 1$, $C_{n+1}^{n+1} = 1$ and the entry of the first and last place of any line of Pascal's triangle is always 1. Now recall that $C_n^k$ is the number of ways of turning on $k$ lamps out of $n$. Consider how many ways there are of turning on $k$ lamps out of $n + 1$. If the last lamp is lit, the others can be lit in $C_n^{k-1}$ ways (as we can turn on $k - 1$ of the remaining lamps). If the last lamp is not lit, we can turn on the others in $C_n^k$ ways. Thus we have shown that $C_{n+1}^k = C_n^{k-1} + C_n^k$. By the inductive assumption $C_n^k$ is the number in the $(n + 1)$st line of Pascal's triangle. But by the definition of Pascal's triangle the next line will consist of the following numbers:

$$1, C_n^0 + C_n^1, C_n^1 + C_n^2, \ldots, C_n^{n-1} + C_n^n, 1.$$

By what has been proved, we can rewrite this line as follows:

$$1, C_{n+1}^1, C_{n+1}^2, \ldots, C_{n+1}^n, 1.$$

This shows that the theorem is true for $n + 1$ as well. Thus the induction is complete.

**66.** We first compute how many ways there are of choosing three objects out of $n$ in a definite order. The first object can be chosen arbitrarily, the second can be chosen from any one of the $n - 1$ remaining objects,

and the third can be chosen from any one of the $n - 2$ objects that remain after the choice of the first two objects. Thus there are $n(n - 1)(n - 2)$ ways. But this method gives us each choice of the objects (with order inessential) exactly six times (we get the objects $A$, $B$, and $C$ in the following six ways: $ABC$, $ACB$, $BAC$, $BCA$, $CAB$, $CBA$). Thus there are $n(n - 1)(n - 2)/6$ ways of choosing 3 objects from $n$ without regard to order.

**67.** The definition of $C_n^k$ tells us that we can choose $k$ objects out of $n$ in $C_n^k$ ways. The solution of our problem will thus consist of an explicit formula for $C_n^k$; we can obtain such a formula in the same way as in Problem 66. But now we shall need to use induction. In Problem 63 we showed that

$$C_n^{k-1} + C_n^k = C_{n-1}^k.$$

We shall prove by induction on $n$ that

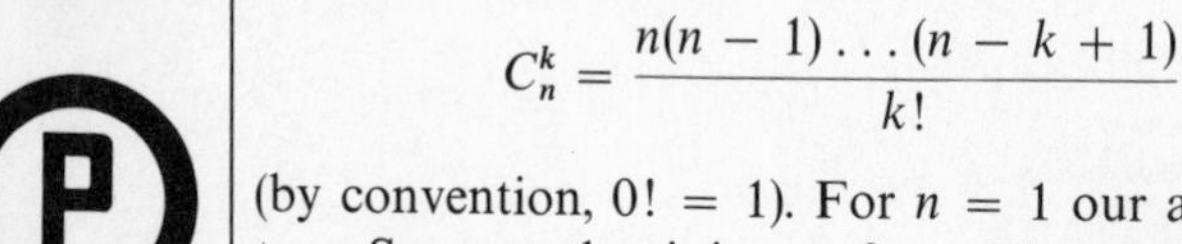

$$C_n^k = \frac{n(n-1)\dots(n-k+1)}{k!}$$

(by convention, $0! = 1$). For $n = 1$ our assertion is true. Suppose that it is true for $n$. Then, by the inductive assumption,

$$\begin{aligned} C_{n+1}^k &= \frac{n(n-1)\dots(n-k+2)}{(k-1)!} + \\ &\quad + \frac{n(n-1)\dots(n-k+1)}{k!} \\ &= \frac{n(n-1)\dots(n-k+2)}{(k-1)!}\left(1 + \frac{n-k+1}{k}\right) \\ &= \frac{n(n-1)\dots(n+1-k+1)}{k!}. \end{aligned}$$

We have proved that our assertion is true for $n + 1$ as well. The induction is thus complete. We can rewrite our formula in the form

$$C_n^k = \frac{n!}{k!(n-k)!}.$$

The number $C_n^k$ is called the *number of combinations of n objects taken k at a time.*

It is the number of ways of choosing $k$ objects out of $n$ without regard to order. There is a special symbol for the number of ways of choosing $k$ objects out of $n$ in a definite order as well: $P_n^k$. We have already used this symbol in the solution of Problem 62; there we proved the formula

$$P_8^5 = C_8^5 \cdot 5! = \frac{8!}{3!}.$$

The analogous formula is valid in the general case as well, that is,

$$P_n^k = C_n^k k! = \frac{n!}{(n-k)!}.$$

Verify this.

The number $P_n^k$ is called the *number of permutations of n objects taken k at a time.*

For $k = n$ the number $P_n^k$ becomes the *number of permutations of n elements* and is sometimes denoted by the symbol $P_n$.

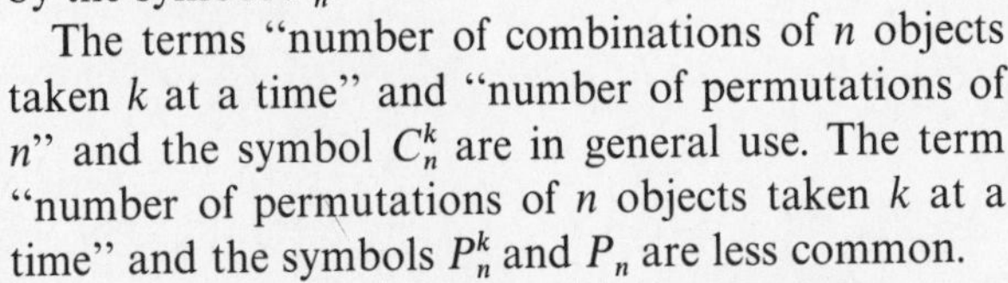

The terms "number of combinations of $n$ objects taken $k$ at a time" and "number of permutations of $n$" and the symbol $C_n^k$ are in general use. The term "number of permutations of $n$ objects taken $k$ at a time" and the symbols $P_n^k$ and $P_n$ are less common.

**70.** Each of the summands that are obtained upon removal of the parentheses in the expression

$$(a + 1)(b + 1)(c + 1)(d + 1)(e + 1)(f + 1)(g + 1)$$

is the product of seven terms (as there are seven pairs of parentheses). Each of these terms is either a letter or the number 1. We are thus required to find the number of products of seven terms each of which can be of either of two forms. Clearly, then, our problem is equivalent to the following:

Find the number of ways of lighting a room using seven lamps, each of which can be either lit or not lit. The answer is $2^7$.

**71.** Reread the solution to Problem 70. We are interested in those terms in which 3 of the factors are letters and 4 are the digit 1. There are as many such terms as there are ways of choosing 3 letters out of 7, that is, $C_7^3 = (7 \cdot 6 \cdot 5)/(1 \cdot 2 \cdot 3) = 35$.

**Remark.** An almost word-for-word repetition of the solution to this problem shows that after multiplying out the expression $(1 + a_1)(1 + a_2) \ldots (1 + a_n)$ there will be $C_n^k$ terms consisting of $k$ letters.

**72.** We expand the expression $(1 + x + y)^{20}$ but do not reduce like terms. Each of the terms obtained in this way contains 20 factors, which can be 1, $x$, or $y$. Our problem is thus equivalent to the following: How many ways are there of lighting 20 traffic lights, each of which can be either red, yellow, or green?

The answer is $3^{20}$ (compare with Problem 70).

**73. Answer**: The coefficient of $x^{17}$ is $(20 \cdot 19 \cdot 18)/2$. The coefficient of $x^{18}$ is 0.

**74.** Reread the remark at the conclusion of the solution of Problem 71. We compare the two expressions

$$(1 + x)^{56} \text{ and } (1 + a_1)(1 + a_2) \ldots (1 + a_{56}).$$

If we set $a_1 = a_2 = \ldots = a_{56} = x$, the second expression becomes identical to the first.

After removal of the parentheses in the expression

$$(1 + a_1)(1 + a_2) \ldots (1 + a_{56}),$$

there will be $C_{56}^8$ terms containing 8 letters. Each of these terms becomes $x^8$ upon substitution of $x$ for $a_1, a_2, \ldots, a_{56}$.

Thus the coefficient of $x^8$ in the expression $(1 + x)^{56}$ is equal to $C_{56}^8$.

Analogously, the coefficient of $x^{48}$ is equal to $C_{56}^{48}$.

**Remark.** The numbers $C_{56}^8$ and $C_{56}^{48}$ are equal. Therefore the coefficients of $x^8$ and $x^{48}$ in the expression $(1 + x)^{56}$ are equal. Similarly, the coefficients of $x^6$ and $x^{50}$ in the expression

$(1 + x)^{56}$ are equal. Likewise, the coefficients of $x^6$ and $x^{50}$ in the expression $(1 + x)^{56}$ are equal. In general, the coefficients in the expression $(1 + x)^n$ of $x^k$ and $x^{n-k}$ ($C_n^k$ and $C_n^{n-k}$, respectively) are equal.

**75.** The proof of the formula

$$(1 + x)^n = C_n^0 + C_n^1 x + C_n^2 x^2 + \dots + \\ + C_n^k x^k + \dots + C_n^n x^n$$

is an almost word-for-word repetition of the solution of Problem 74.

**76.** We carry out the proof by induction. For $n = 10$ the assertion is true: $(1 + x)^1 = 1 + 1 \cdot x$, and so the binomial coefficients coincide with the numbers in the second line of Pascal's triangle.

Assume that the assertion is true for $n - 1$, that is, that

$$(1 + x)^{n-1} = a_0 + a_1 x + \dots + a_{k-1} x^{k-1} + a_k x^k + \\ + \dots + a_{n-1} x^{n-1},$$

where $a_0, a_1, \dots, a_{k-1}, a_k, \dots, a_{n-1}$ are the numbers in the $n$th line of Pascal's triangle.

Let us prove that it is then also valid for $n$. In other words, it is necessary to prove that

$$(1 + x)^n = b_0 + b_1 x + \dots + b_k x^k + \dots + b_n x^n,$$

where $b_0, b_1, \dots, b_k, \dots, b_n$ are the numbers in the $(n + 1)$st line of Pascal's triangle.

By the definition of Pascal's triangle we have

$$b_0 = 1, \quad b_1 = a_0 + a_1, \dots, \quad b_k = a_{k-1} + a_k, \dots, \\ b_{n-1} = a_{n-2} + a_{n-1}, \quad b_n = 1 (= a_{n-1}).$$

Thus, we must show that

$$(1 + x)^n = 1 + (a_0 + a_1)x + \dots + (a_{k-1} + a_k)x^k + \\ + \dots + (a_{n-2} + 1)x^{n-1} + 1 \cdot x_1^n.$$

But this is obvious:

$$(1 + x)^n = (1 - x)^{n-1}(1 + x) = (a_0 + a_1x + \ldots + a_kx^k + \ldots + a_{n-1}x^{n-1})(1 + x)$$
$$= a_0 + a_1x + \ldots + a_kx^k + \ldots + a_{n-1}x^{n-1} + a_0x + a_1x^2 + \ldots + a_{k-1}x^k + \ldots + a_{n-2}x^{n-1} + a_{n-1}x^n$$
$$= a_0 + (a_0 + a_1)x + \ldots + (a_{k-1} + a_k)x^k + \ldots + (a_{n-2} + a_{n-1})x^{n-1} + a_{n-1}x^n.$$

**78.** Since $(a + b)^n = a^n(1 + b/a)^n$, the term containing $a^k$ is equal to

$$C_n^k a^k b^{n-k}.$$

**79.** First we find all of the terms in $x^k$. We write

$$(x + y + z)^n = [x + (y + z)]^n = \ldots + C_n^k x^k (y + z)^{n-k} + \ldots.$$

In the expression $(y + z)^{n-k}$ we now find the term in $y^l$:

$$(y + z)^{n-k} = \ldots + C_{n-k}^l y^l z^{n-k-l} + \ldots.$$

Thus it is clear that the term containing $x^k y^l$ will be

$$C_n^k C_{n-k}^l x^k y^l z^{n-k-l}.$$

The expression $C_n^k \cdot C_{n-k}^l$ can be transformed slightly:

$$C_n^k \cdot C_{n-k} = \frac{n!}{k!(n-k)!} \cdot \frac{(n-k)!}{l!(n-k-l)!} = \frac{n!}{k!l!(n-k-l)!}.$$

**80.** If in the identity

$$(1 + x)^n = C_n^0 + C_n^1 x + C_n^2 x^2 + \ldots + C_n^{n-1} x^{n-1} + C_n^n x^n$$

we set $x = 1$, we get

$$2^n = C_n^0 + C_n^1 + C_n^2 + \ldots + C_n^{n-1} + C_n^n.$$

For $x = -1$ we get

$$0 = C_n^0 - C_n^1 + C_n^2 + \ldots + (-1)^n C_n^n.$$

Adding and subtracting these equalities term by term, we get

$$C_n^0 + C_n^2 + C_n^4 + \ldots = C_n^1 + C_n^3 + C_n^5 + \ldots = 2^{n-1}.$$

**81.** If we expand the expression $(1 + x - 3x^2)^{1965}$ and reduce like terms, we get a polynomial of the form $a_0 + a_1x + a_2x^2 + a_3x^3 + \ldots$. We note that the sum of its coefficients is equal to the value of the polynomial for $x = 1$:

$$a_0 + a_1 \cdot 1 + a_2 \cdot 1^2 + a_3 \cdot 1^3 + \ldots = a_0 + a_1 + a_2 + a_3 + \ldots.$$

Of course, it is in fact unnecessary to multiply and reduce like terms. It suffices to insert $x = 1$ in the initial expression

$$(1 + 1 - 3 \cdot 1^2)^{1965} = (-1)^{1965} = -1.$$

Thus, $a_0 + a_1 + a_2 + a_3 + \ldots = -1$.

**86. Answer:** $C_n^2 - n$. **Solution:** Take all pairs of distinct vertices of the polygon (there are $C_n^2$ such pairs) and join the points of each pair by line segments. We obtain $C_n^2$ segments. Of them, $n$ will be sides and the remainder will be diagonals; that is, there will be $C_n^2 - n$ diagonals.

**91.** We rewrite the conditions of the problem in a more compact form:

| E | G | F | EG | GF | FE | EGF |
|---|---|---|---|---|---|---|
| 6 | 6 | 7 | 4 | 3 | 2 | 1 |

Suppose that the person knowing all three languages leaves the room; then none of the remaining occupants of the room will know more than two languages and we shall have the following problem:

| E | G | F | EG | GF | FE | EGF |
|---|---|---|---|---|---|---|
| 5 | 5 | 6 | 3 | 2 | 1 | 0 |

Suppose now that the three persons knowing both English and German leave the room; then the number of people who know the other pairs of languages remains the same (since there is no one in the room who knows all three languages):

| E | G | F | EG | GF | FE | EGF |
|---|---|---|---|---|---|---|
| 2 | 2 | 6 | 0 | 2 | 1 | 0 |

If the two people knowing both German and French and the person who knows both French and English leave, we get the problem:

| E | G | F | EG | GF | FE | EGF |
|---|---|---|---|---|---|---|
| 1 | 0 | 3 | 0 | 0 | 0 | 0 |

But this problem is easy to solve: One person knowing only English and three people knowing only French are left in the room; and so there are four people in all. But seven people left the room; thus, there were initially eleven people in the room. Only one person knows only English.

**92.** We add up the number of readers who have read each selection of $k$ books for each such selection. We denote the sum obtained by $S_k$. We shall prove that

$$S = S_1 - S_2 + S_3 - S_4 + \dots + (-1)^{n-1}S_n$$

is the number of readers in the library.

Let us consider a reader who has read exactly $k$ books and examine the contribution that he makes to each of the terms making up $S$. Without loss of generality we can assume that our reader has read the first $k$ books. Our reader contributes $k(= C_k^1)$ to the summand $S_1$, because he has read each of $k$ books. Our reader's contribution to the summand $S_2$ is $C_k^2$ since he has read each pair of books from the first $k$. Analogously, his contribution to $S_m$ is $C_k^m$ for $m \leq k$, and $O$ for $m > k$. Thus, the contribution of our reader to the sum $S$ is

$$C_k^1 - C_k^2 + C_k^3 - \dots + (-1)^{k-1}C_k^k.$$

But we know that $-C_k^0 + C_k^1 - C_k^2 + C_k^3 + \dots + (-1)^{k-1}C_k^k = 0$ and $C_k^0 = 1$. Thus our reader contributes 1 to the sum $S$. This argument applies to any reader; consequently, the number of readers in the library is equal to $S$, which is what we were required to prove.

**Remark.** It is not difficult to see that Problem 91 is a special case of Problem 92. However, the method of solution of Problem 91 can be used for the solution of Problem 92 as well.

**93.** We consider *five* sets of telephone numbers:

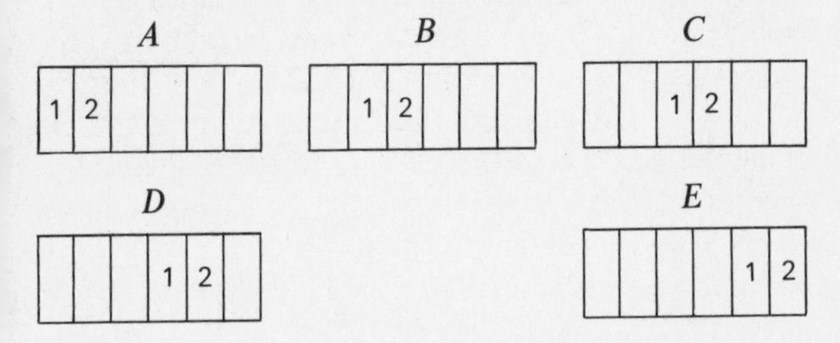


(the set $B$, for example, consists of the numbers having 12 in the second and third places and arbitrary digits in the other places). Each of these sets contains $10^4$ numbers. The sets $A$ and $B$ do not have a single number in common. The sets $A$ and $C$ have $10^2$ numbers in common—the numbers of the form

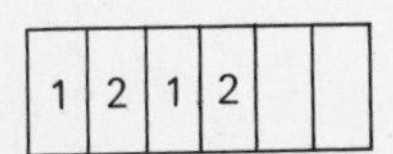


It is easy to see that there are $6 \cdot 10^2$ numbers which belong simultaneously to two or more sets, and only one number (12 12 12) which is a member of three of the sets. Now, either by the method of solution of Problem 91 or by that of Problem 92, we can easily show that the desired number is equal to $5 \cdot 10^4 - 6 \cdot 10^2 + 1 = 49{,}401$.

**95.** If it does not rain even once on a particular day, both students mark down a "+." If it is raining during both observations, both write "−." If it is raining in the morning but not in the afternoon or evening, the first student marks down "−" and the second, "+."

The case + − is impossible because the first student enters a "+" only if it does not rain at all. But then the second student must enter a "+."

**96.** If the first student enters a "+," it could not have been raining during any of the observations. Then the other two students also mark down a "+." The total entry will be + + +. If the first student enters a "−," and the third a "+," it rained during exactly one of the three observations. Then the second student enters a "+." The total entry will thus be "− + +." If the first and third student enter a "−," it rained either two or three times. In the first case the second student enters a "+," in the second case, a "−." The total entry will be "− + −" or "− − −." Since these are all the cases that are possible, the other entries will not occur.

Another method of solution is the following. It can either be clear or rain once, twice, or three times. We obtain our answer from the table:

| Number of Times That It Rains / Entries | 0 | 1 | 2 | 3 |
|---|---|---|---|---|
| First Student | + | − | − | − |
| Second Student | + | + | + | − |
| Third Student | + | + | − | − |

**97.** (a) Let $A$ be the shortest of the tall people and $B$ the tallest of the short people. We compare them with $C$, the person in $A$'s row and $B$'s column. Since $A$ is the tallest person in his row, he is taller than $C$, and since $B$ is the shortest person in his column, he is shorter than

$C$. Thus, $A$ is taller than $B$.

(b) Let us consider the two arrangements (Fig. 34).

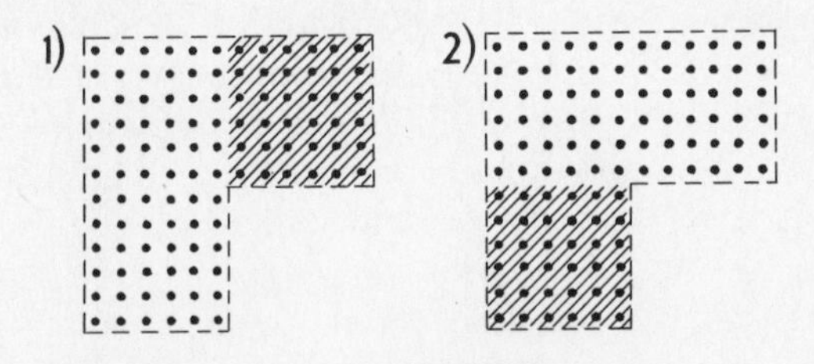


Fig. 34

In the first case, if the people in the shaded positions are taller than the others, the shortest of the tall people will be taller than the tallest of the short people. In the second case, if the people in the shaded positions are shorter, the shortest of the tall will be shorter (verify this!).

Consider why in the second of these cases it is impossible to apply the same argument as in the solution of Problem (a).

**99.** Suppose that each student solves only one problem in a test but in such a way that each problem is solved by some student. Then the test will be difficult in the sense of (a) and easy in the sense of (b).

**100.** The examples $3 + 4 = 7, 2 + 7 = 9$ show that Theorems 2, 3, 4, and 5 are false. Theorem 1 is obviously true. Theorem 6 is easily proved by contradiction.

**101.** Theorem 8 is easily proved by contradiction (if Theorem 1 is true). For suppose that Theorem 8 is false. Then $\bar{\mathbf{B}}$ will not imply $\bar{\mathbf{A}}$, that is, the case where $\bar{\mathbf{B}}$ but not $\bar{\mathbf{A}}$ is satisfied is possible. But this means that **B** is not satisfied while **A** is. But by Theorem 1 this case is impossible.

For Theorems 4 and 5 we can find examples of propositions that satisfy these theorems and examples of propositions that do not. The most obvious examples can be constructed as follows. Suppose that $A$ and $B$ are two sets of points in the plane. For **A** take the proposition "The point $M$ belongs to the set $A$," and for **B** take the proposition "The point $M$ belongs to the set $B$." Then Theorem 1 states that the set $A$ is entirely contained in the set $B$; Theorem 2 states that the complement of $A$[1] is contained in $B$; Theorem 3 states that $A$ is contained in the complement of $B$, and so on. Formulate the remaining theorems by yourself and draw examples of sets $A$ and $B$ for which these theorems are true and examples for which they are false.

The examples we need can be obtained from an examination of Fig. 35.

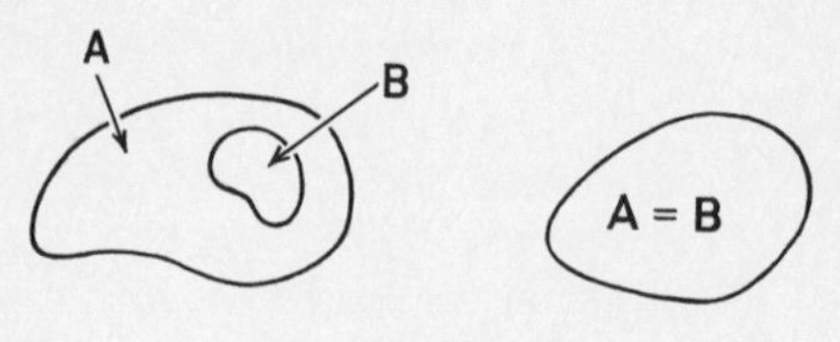


Fig. 35

Analyze these cases by yourself and convince yourself that Theorems 4 and 5 can be either true or false.

It would, of course, be possible to find other examples (such as Theorems 1 and 6 of Problem 100 or theorems that you know from geometry and algebra). We shall now show that Theorems 2, 3, 6, and 7 are false.

If Theorem 2 were true, we would have the following scheme:

$$\mathbf{A} \xrightarrow{\text{Theorem 1}} \mathbf{B} \xleftarrow{\text{Theorem 2}} \bar{\mathbf{A}}$$

(if **A** is true, **B** is true by Theorem 1, and if **A** is false, **B** is true by Theorem 2). Thus, assertion **B** would always have to be true, but we have agreed not to consider such propositions.

If Theorem 3 is true, then

$$\bar{\mathbf{B}} \xleftarrow{\text{Theorem 3}} \mathbf{A} \xrightarrow{\text{Theorem 1}} \mathbf{B}.$$

[1]The complement of a set $A$ in the plane is the set of all points of the plane that are not members of $A$. The complement of the upper half-plane (including the $x$-axis), for example, is the lower half-plane (excluding the points of the $x$-axis).

Thus, whenever Proposition **A** is true, the mutually exclusive Propositions **B** and $\bar{\mathbf{B}}$ are also true, which is impossible. Thus **A** is always false. But we are not considering such propositions.

If Theorem 6 is true, then

$$\bar{\mathbf{B}} \xrightarrow{\text{Theorem 6}} \mathbf{A} \xrightarrow{\text{Theorem 1}} \mathbf{B}.$$

We see that if **B** is false, we can prove that it is true. But this can occur only if **B** is always true.

If Theorem 7 is true,

$$\mathbf{A} \xrightarrow{\text{Theorem 1}} \mathbf{B} \xrightarrow{\text{Theorem 7}} \bar{\mathbf{A}}.$$

In other words, if **A** is true, then it is false. Thus, **A** is always false.

**102.** (a) Suppose first that $x \geq 0$. Then $|x| = x$ and we get the equation $3x = 3$, whence $x = 1$. Now suppose that $x \leq 0$: then $|x| = -x$, and we get the equation $-x = 3$, whence $x = -3$. The answer is thus $x_1 = 1, x_2 = -3$.

(b) For $x \geq 0$ we get the equation $x^2 + 3x - 4 = 0$, whence $x_1 = 1$, $x_2 = 4$. The condition $x \geq 0$ is satisfied only by the first root.

For $x \leq 0$ we get the equation $x^2 - 3x - 4 = 0$, whence $x_1 = -1$, $x_2 = 4$. The condition $x \leq 0$ is satisfied only by the first root. Thus the answer is $x_1 = 1, x_2 = -1$.

(c) Suppose that $x \leq -\frac{1}{2}$. Then $|2x + 1| = -(2x + 1)$, $|2x - 1| = -(2x - 1)$. We get the equation $-(2x + 1) - (2x - 1) = 2$, whence $x = -\frac{1}{2}$.
$- 4 = 0$, whence $x_1 = 1$, $x_2 = -4$. The condition

Suppose that $-\frac{1}{2} \leq x \leq \frac{1}{2}$. Then $|2x + 1| = 2x + 1$, $|2x - 1| = -(2x - 1)$. We get the equation $(2x + 1) - (2x - 1) = 2$, which is an identity. Thus, all of the numbers in the segment $[-\frac{1}{2}, \frac{1}{2}]$ are solutions of our equation.

Suppose now that $x \geq \frac{1}{2}$. Then $|2x + 1| = 2x + 1$ and $|2x - 1| = 2x - 1$. We get the equation $(2x + 1) + (2x - 1) = 2$, whence $x = \frac{1}{2}$. Thus the answer is that all numbers of the segment $[-\frac{1}{2}, \frac{1}{2}]$ are solutions.

**103.** (a) Suppose first that the numbers $x$ and $y$ are positive. Then $|x| = x$, $|y| = y$, $|x + y| = x + y$. The inequality under consideration is then an equality.

If $x$ is positive and $y$ negative, we must consider the cases $x + y \geq 0$ and $x + y \leq 0$ separately. In the first case $|x| = x$, $|y| = -y$, $|x + y| = x + y$. The inequality takes the form $x + y \leq x - y$. This is true since $y$ is negative.

In the second case $|x| = x$, $|y| = -y$, $|x + y| = -(x + y)$. The inequality takes the form

$$-x - y \leq x - y.$$

This is true, since $x$ is positive.

The remaining cases can be obtained from those already analyzed if we interchange the signs of both $x$ and $y$. Since in this process $|x|$, $|y|$, and $|x + y|$ do not change, the inequality remains valid.

(b) We could, as in the solution to Problem 9a, consider separately all arrangements of the numbers $x$, $y$, and $x - y$ on the number axis, but we prefer to use the fact that the inequality $|x + y| \leq |x| + |y|$ has already been proved. Denote $x - y$ by $z$. Then $x = y + z$. Since $|y + z| \leq |y| + |z|$, $|x| \leq |y| + |x - y|$, which is what we wanted to prove.

(c) Here again we can obtain the solution by considering the various cases, but it is simpler to deduce this inequality from those already proved. We note that if $|x| \geq |y|$, our inequality coincides with the inequality of Problem 103b. If $|x| < |y|$, our inequality takes the form $|x - y| \geq |y| - |x|$ or $|y - x| \geq |y| - |x|$. But again this is the inequality of Problem 103b if we interchange $x$ and $y$.

**Remark.** We can offer a more intuitive solution by using the fact that the quantity $|x|$ is equal to the distance between the points $x$ and 0 on the number axis, and that the quantity $|x - y|$ is the distance between the points $x$ and $y$. See the first volume of our "Library," *The Method of Coordinates.*

**104.** (a) The inequality $\sqrt[n]{1000} < 1.001$ can be rewritten as follows: $1000 < (1 + 0.001)^n$. We write out the right side using the binomial formula. We get

$$(1 + 0.001)^n = 1 + \frac{n}{1000} + \frac{n(n-1)}{2 \cdot 1000^2} + \ldots + \frac{1}{1000^n}.$$

From this it is clear that for $n > 1$ the quantity $(1 + 0.001)^n$ is always greater than $1 + 1000/n$. Therefore for sufficiently large $n$, for example, for $n = 1{,}000{,}000$, we will have $(1 + 0.001)^n > 1000$. Thus for $n = 1{,}000{,}000$ the initial inequality $\sqrt[n]{1000} < 1.001$ is satisfied.

(b) As in the solution of Problem 104a, we rewrite the inequality in the form $n < (1 + 0.001)^n$ and expand the right side, using the binomial formula. From this formula it follows that for $n > 2$ the inequality $(1 + 0.001)^n > 1 + n/1000 + n(n-1)/2 \cdot 1000^2$ is valid. But for sufficiently large $n$ this expression is greater than $n$. In fact, if $n - 1 > 2 \cdot 1000^2$, the last summand will exceed $n$. Consequently, for $n = 2 \cdot 1000^2 + 2$ the inequality $(1.001)^n > n$ and thus the initial inequality $\sqrt[n]{n} < 1.001$ as well are satisfied.

(c) We use the identity $\sqrt{n+1} - \sqrt{n} = 1/(\sqrt{n+1} + \sqrt{n})$. From this it is clear that the quantity $\sqrt{n+1} - \sqrt{n}$ is in any case not greater than $1/2\sqrt{n}$. Therefore for $n > 25$ the inequality

$$\sqrt{n+1} - \sqrt{n} < \frac{1}{2\sqrt{n}} < \frac{1}{10}$$

is satisfied.

(d) No such number $n$ exists. In fact, if $\sqrt{n^2 + n} - n < 0.1$, $\sqrt{n^2 + n} < n + 0.1$ and $n^2 + n < (n + 0.1)^2 = n^2 + n/5 + 0.01$. The latter inequality is obviously not satisfied for any natural number $n$ since $n > n/5 + 0.01$.

**105.** Let us consider how the expression

$$\left| \frac{(k^3 - 2k + 1)}{(k^4 - 3)} \right|$$

behaves for values of $k$ whose absolute value is large. It is clear that $k^3$ is the most important term in the numerator, and that the term $k^4$ plays the main role in the denominator. Therefore, we may expect that for large values of $k$ our expression will be approximately equal to $|k^3/k^4| = 1/|k|$. We now investigate the extent to which the exact value of our expression differs from this approximate value. We make the transformation:

$$\left|\frac{k^3 - 2k + 1}{k^4 - 3}\right| = \frac{\left|k^3\left(1 - \dfrac{2}{k^3} + \dfrac{1}{k^3}\right)\right|}{k^4\left(1 - \dfrac{3}{k^4}\right)}$$

$$= \frac{1}{|k|}\frac{\left|1 - \dfrac{2}{k^2} + \dfrac{1}{k^3}\right|}{\left|1 - \dfrac{3}{k^4}\right|}.$$

Suppose $|k| \geq 2$; then

$$\left|1 - \frac{2}{k^2} + \frac{1}{k^3}\right| \leq 1 + \frac{2}{k^2} + \frac{1}{|k|^3} \leq 1 + \frac{1}{2} + \frac{1}{8} < 2,$$

$$\left|1 - \frac{3}{k^4}\right| \geq 1 - \frac{3}{k^4} \geq 1 - \frac{3}{16} > \frac{1}{2}.$$

Therefore for $|k| \geq 2$ the inequality

$$\frac{1}{|k|}\frac{\left|1 - \dfrac{2}{k^2} + \dfrac{1}{k^3}\right|}{\left|1 - \dfrac{3}{k^4}\right|} < \frac{1}{2}\cdot\frac{2}{\frac{1}{2}} = 2.$$

is valid.

Thus, for $|k| \geq 2$ our expression does not exceed 2. All that remains is to consider what values it assumes for $k = -1$, 0, and 1. These values are 1, $\frac{1}{3}$, and 0, respectively.

Thus the answer is that a suitable number $C$ exists. We can, for example, take

$$C = 2.$$

**Remark.** Using the same method, we can get a more exact estimate of our expression and convince ourselves that for $k = -1$ it assumes its greatest value, equal to 1. Carry this out by yourself.

**106.** Our expression is a product of two numbers: $k$ and $\sin k$. The first of these numbers can be chosen arbitrarily large. If for some $k$ the second number is not too small, the entire product will be a large number. We demand, for example, that $\sin k$ be greater than $\frac{1}{2}$. The set of points $x$ for which $\sin x > \frac{1}{2}$ consists of an infinite number of intervals of the form

$$2\pi n + \frac{\pi}{6} < x < 2\pi n + \frac{5\pi}{6},$$

where $n$ is an arbitrary integer (Fig. 36). The length of each of these intervals is equal to $2\pi/3$. Since this number is greater than 1, each of these intervals contains at least one integer. From this it follows that for any number $C$ there exist infinitely many numbers for which $k \sin k > C$. Indeed, for each number $k$ lying in the intervals just mentioned, the inequality $\sin k > \frac{1}{2}$ is satisfied. Therefore if the natural number $k$ is greater than $2C$ and lies within one of the intervals shown, $k \sin k > C$. Clearly, there are infinitely many such numbers.

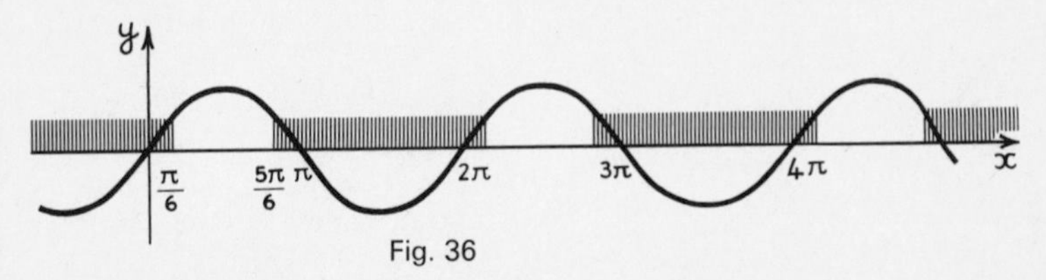


Fig. 36

**107.** (a) Suppose that the segment $[a, b]$ is a trap. That is, at most a finite number of terms of the sequence lie outside the segment $[a, b]$. If this segment is not a trough it contains only a finite number of terms of the sequence. But a sequence has an infinite numbers of terms. This contradiction shows that the segment $[a, b]$ must be a trough.

(b) The sequence (b) and the segment B of

Problem 108. The odd-numbered terms of the sequence lie within the segment, and the even-numbered ones do not. Thus the segment is a trough but not a trap.

**109.** (a) The sequence

$$1, 3, \tfrac{1}{2}, 2\tfrac{1}{2}, \ldots, \tfrac{1}{n}, 2\tfrac{1}{n}, \ldots.$$

(b) There is no such sequence. In fact, suppose that for a certain sequence the segment [0, 1] is a trap. Then only a finite number of terms of the sequence can lie outside this segment. Thus, only finitely many terms of the sequence can lie inside the segment [2, 3]. Thus the segment [2, 3] is not a trough and all the more so (see Problem 107a) is not a trap for the sequence.

**110.** (a) Any segment of length 1 fails to intersect at least one of the segments [0, 1] and [9, 10]. Suppose, for example, that a segment has no point in common with the segment [0, 1]. If this segment were a trap, there would be only a finite number of terms of the sequence lying outside this segment and, in particular, inside the segment [0, 1]. But this contradicts the assumption that the segment [0, 1] is a trough.

(b) Consider the two sequences:

1. $1, 9, \tfrac{1}{2}, 9\tfrac{1}{2}, \tfrac{1}{3}, 9\tfrac{2}{3}, \ldots, \tfrac{1}{n}, 9\dfrac{n-1}{n}, \ldots,$

2. $0, 10, \tfrac{1}{2}, 9\tfrac{1}{2}, \tfrac{2}{3}, 9\tfrac{1}{3}, \ldots, \dfrac{n-1}{n}, 9, \ldots.$

The first sequence does not have a trap of length 9. (Verify that the segments $[0, \tfrac{1}{3}]$ and $[9\tfrac{2}{3}, 10]$ are troughs for this sequence. This will imply, as in the solution to Problem 110a, that there is no trap of length 9.) The segment $[\tfrac{1}{2}, 9\tfrac{1}{2}]$ is a trap for the second sequence since each term beginning with the third belongs to this segment.

**111.** (a) The sequence $1, 2, 3, \ldots, n, \ldots$ has no troughs since no more than $l + 1$ terms of the sequence are contained in a segment of length $l$.

(b) We construct a sequence whose terms include all rational numbers. Since any segment contains an infinite number of rational numbers, any segment will be a trough for this sequence. To construct the required sequence, we draw a lattice in the plane as shown in Figure 37 and move along the lattice in the direction of the arrows.

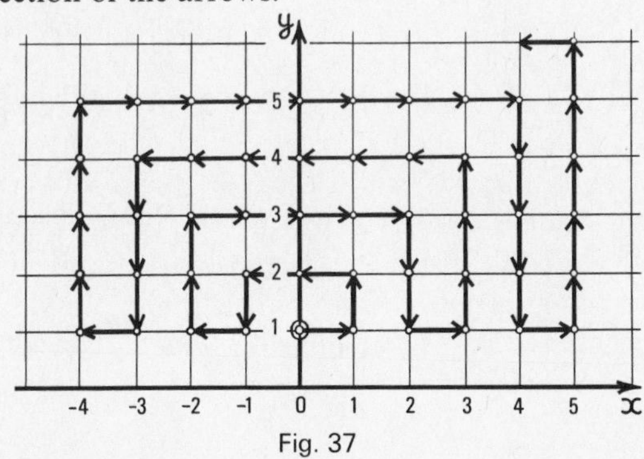

Fig. 37

Each time that we pass a vertex of the lattice, we write down a term of the sequence. If the lattice point has the coordinates $x$, $y$ (as is evident from the sketch, $x$ and $y$ are integers and $y > 0$), the corresponding term of the sequence is $x/y$. In this way we obtain the sequence

$$\tfrac{0}{1}, \tfrac{1}{1}, \tfrac{1}{2}, \tfrac{0}{2}, -\tfrac{1}{2}, -\tfrac{1}{1}, -\tfrac{2}{1}, -\tfrac{2}{2}, -\tfrac{2}{3}, \ldots.$$

We shall show that each rational number is assumed by some term of this sequence. In fact, each rational number is by definition of the form $p/q$, where $p$ and $q$ are integers and $q > 0$. Movement along the lattice in the prescribed manner eventually gets us to the point with coordinates $p$, $q$. The corresponding term of the sequence is then $p/q$.

Actually, our sequence assumes each rational number an infinite number of times, since each rational number has an infinite number of representations in the form $p/q$ (for example, $\frac{2}{3} = \frac{4}{6} = \frac{6}{9} = \frac{8}{12} = \ldots$). It is, however, not quite so simple to give for each rational number the index of the corresponding term of the sequence, explicitly. Try, for example, to determine the position of the numbers $1\frac{1}{3}$, $-\frac{5}{7}$, and 10 in the preceding sequence. And what is its 100th term?

**113.** (a) We consider a segment whose center is the point $a$. We denote the length of this segment by $2\varepsilon$. By the definition of limit there exists a number $k$ such that for all $n > k$ the inequality $|x_n - a| < \varepsilon$ is valid. This inequality asserts that the point $x_n$ is contained in our segment. Thus, not more than $k$ terms of the sequence lie outside of our segment. It is thus a trap.

(b) Suppose that we are given an arbitrary positive number $\varepsilon$. We consider a segment whose center is $a$ having length less than $2\varepsilon$. By assumption, this segment is a trap. Therefore, only finitely many terms of the sequence lie outside it. Suppose that the $k$th position is the last in which such a term appears. (If no terms of the sequence lie outside the segment, we set $k = 0$.) Then for any $n > k$ the term $x_n$ lies on our segment; that is, the inequality $|x_n - a| < \varepsilon$ is satisfied. We have thus proved that $a$ is the limit of the sequence $\{x_n\}$.

**114.** (a) Suppose that we are given a segment whose center is the point $a$. We denote the length of this segment by $2\varepsilon$. By the definition of a limit there exists a number $k$ for which $n > k$ implies that the inequality $|x_n - a| < \varepsilon$ is satisfied. This means that all terms of the sequence after the $k$th lie on the given segment.

Suppose now that we are given a segment not containing the point $a$. We denote by $\varepsilon$ the distance of the point $a$ to the nearer end point of the segment. By the definition of a limit there exists a number $k$ such that for $n > k$ the inequality $|x_n - a| < \varepsilon$ is satisfied. This implies that for $n > k$ the term $x_n$ lies closer to the point $a$ than does the nearer end point of the segment. Therefore, all of the terms after the $k$th lie outside of the segment. This implies that it is impossible for an infinite set of terms of the sequence to lie on the given segment.

(b) For the sequence $1, \frac{1}{2}, 3, \frac{1}{4}, \ldots, n^{(-1)^{n-1}}, \ldots$ any segment having center 0 is a trough. In fact, the subsequence of terms in the even-numbered positions

$$\frac{1}{2}, \frac{1}{4}, \frac{1}{6}, \ldots, \frac{1}{2n}, \ldots$$

clearly converges to zero. Therefore any segment having center 0 contains an infinite number of terms of this subsequence (compare the solution of Problem 114a). Thus, this segment is a trough for the initial sequence. We now show that no segment not containing the point 0 is a trough. Indeed, for the subsequence of terms in the even positions, this segment is not a trough (compare the solution of Problem 114a); for the subsequence of terms in the odd positions

$$1, 3, 5, 7, \ldots,$$

no segment at all can be a trough (prove this by yourself). Therefore on each segment not containing the point 0, there will be only finitely many terms in even-numbered positions and only finitely many in odd-numbered positions. This implies that this segment contains only finitely many terms of the sequence and consequently is not a trough.

Thus, the sequence $x_n = n^{(-1)^{n-1}}$ and the number 0 satisfy the conditions of Problem 114b. But the number 0 is not a limit of the sequence. For otherwise, the condition $|x_n| < 1$ would be satisfied for all terms of the sequence for which $n$ is sufficiently large. However, no term in the odd-numbered positions satisfies this inequality.

**Remark.** The sequence

$$x_n = n^{(-1)^{n-1}}$$

can be represented as the union of two simpler sequences:

$$1, 3, 5, 7, \ldots, 2n-1, \ldots \quad \text{and} \quad \tfrac{1}{2}, \tfrac{1}{4}, \tfrac{1}{6}, \ldots, \tfrac{1}{2n}, \ldots.$$

This method of constructing a sequence is frequently useful. For example, to construct an example of a sequence for which each of two given segments is a trough (see Problem 109b), we can join together two sequences the first segment being a trough for the first sequence and the second segment being a trough for the second sequence.

**115.** In the solution to Problem 114a we proved that if $x_n \to a$ as $n \to \infty$, no segment which does not contain the point $a$ is a trough. The required assertion can easily be deduced from this by the method of proof by contradiction. Carry this out by yourself.

**116.** To prove that the sequence $\{x_n\}$ has the number $a$ as limit, we must show that for each positive number $\varepsilon$ we can find a number $k$ such that for $n > k$ the inequality $|x_n - a| < \varepsilon$ will be valid. It is frequently possible to find an explicit formula expressing $k$ as a function of $\varepsilon$.

(a)

$$x_n = \frac{(-1)^{n-1}}{n}, \qquad \lim_{n\to\infty} x_n = 0.$$

Since $|x_n| = 1/n$, we can take $1/\varepsilon$ as $k$. For when $n > 1/\varepsilon$ the inequality $|x_n| = 1/n < \varepsilon$ is satisfied.

**Remark.** The number $k$ is not uniquely determined by $\varepsilon$. In the above example we could have taken $k = 2/\varepsilon$ or $(1/\varepsilon) + 1$ (and in general, any number greater than $1/\varepsilon$). It is frequently convenient not to take the most "economical" value for $k$ (which can be very complicated to express in terms of $\varepsilon$), but a simpler expression. For example, to prove that

$$\lim_{n\to\infty} \frac{1}{n^3 + 0.6n + 3.2} = 0,$$

we can use the fact that $x_n \leq 1/n^3 \leq 1/n$ and take $1/\varepsilon$ for $k$. This is much more convenient than the most economical value

$$k = \sqrt[3]{1.6 + \sqrt{2.552 - \frac{1.6}{\varepsilon} + \frac{1}{4\varepsilon^2}}} + \sqrt[3]{1.6 - \sqrt{2.552 - \frac{1.6}{\varepsilon} + \frac{1}{4\varepsilon^2}}},$$

which is obtained by solving the cubic equation

$$n^3 + 0.6n + 3.2 = 1/\varepsilon.$$

There are also cases where it is impossible to find an explicit formula for the most economical value of $k$, but an appropriate larger value can easily be found.

(b)

$$x_n = \frac{3^n - 1}{3^n} = 1 - \frac{1}{3^n}; \quad \lim_{n\to\infty} x_n = 1.$$

Since $|x_n - 1| = 1/3^n$, we can take $\log_3 (1/\varepsilon)$ as $k$.

(c) Using the formula for the sum of a geometric progression, we find $x_n = 1 + \frac{1}{2} + \ldots + (1/2^n) = 2 - (1/2^n)$. From this it is evident that $\lim_{n\to\infty} x_n = 2$ and that for $k$ we can take $\log_2 (1/\varepsilon)$.

(d) There is no limit because for each number $a$ and each $\varepsilon$ the inequality $|x_n - a| < \varepsilon$ is satisfied for only finitely many $n$.

(e) $\lim_{n\to\infty} x_n = 1$; since $|x_n - 1|$ in this case is identically equal to zero, the inequality $|x_n - 1| < \varepsilon$ is satisfied for each positive $\varepsilon$ and all $n$.

(f) We consider the case of even and of odd $n$ separately. For even $n = 2m$ we have $x_n = 1/m$. Therefore for $m > 1/\varepsilon$ (that is, for $n > 2/\varepsilon$) the inequality $|x_n| < \varepsilon$ is satisfied. For odd $n$ the term $x_n$ is equal to zero, and therefore the inequality $|x_n| < \varepsilon$ is satisfied for all odd $n$. Thus, $\lim_{n\to\infty} x_n = 0$ and for $k$ we can take the number $2/\varepsilon$.

(g) By the formula for the sum of a geometric progression we have

$$0.\underbrace{22\ldots2}_{n} = \frac{2}{10} + \frac{2}{10^2} + \ldots + \frac{2}{10^n} = \frac{2}{9}\left(1 - \frac{1}{10^n}\right).$$

From this we get $\lim_{n\to\infty} x_n = \frac{2}{9}$ and for $k$ we can take log $(\frac{2}{9}\varepsilon)$ (or simply $\log (1/\varepsilon)$ – see the remark on the solution of Example 116a).

(h) There is no limit. In fact, if $n = 180m$, $x_n = 0$, and if $n = 90 + 360m$, $x_n = 1$. If the sequence had the limit $a$, the inequality $|x_n - a| < \frac{1}{4}$ would be satisfied for all $n > n_0$ for some integer $n_0$. From this, in turn, it would follow that for $n_1 > n_0$ and $n_2 > n_0$ the inequality

$$|x_{n_1} - x_{n_2}| \leq |x_{n_1} - a| + |a - x_{n_2}| < \tfrac{1}{4} + \tfrac{1}{4} = \tfrac{1}{2}$$

would be satisfied.

But our sequence has the property that for any $n$ there exist terms $x_{n'}$ and $x_{n''}$ with $n'$ and $n'' > n$ such that $|x_{n'} - x_{n''}| > \frac{1}{2}$.

(i) Since $|\cos n°| \leq 1$, $|x_n| \leq 1/n$. From this it is clear that $\lim_{n\to\infty} x_n = 0$ and that for $k$ we can take $1/\varepsilon$.

(j) There is no limit. If the number $a$ were the limit, the inequality $|x_n - a| < \frac{1}{2}$ would be satisfied for all sufficiently large $n$. Then $|x_n - x_{n+1}| \leq |x_n - a| + |a - x_{n+1}| < \frac{1}{2} + \frac{1}{2} = 1$. But the difference between any two consecutive terms of our sequence is greater than 1.

**117.** Suppose that two distinct numbers $a$ and $b$ are limits of the same sequence $\{x_n\}$. We denote the distance between these points by $2\varepsilon$. If $a$ is the limit of the sequence $\{x_n\}$, then there exists a number $k_1$ such that for $n > k_1$ the inequality $|x_n - a| < \varepsilon$ will be satisfied. Analogously, if $b$ is a limit of the sequence $\{x_n\}$ there exists a number $k_2$ such that for $n > k_2$, $|x_n - b| < \varepsilon$. Consequently, if the number $n$ is greater than both $k_1$ and $k_2$, both inequalities $|x_n - a| < \varepsilon$ and $|x_n - b| < \varepsilon$ are satisfied. But this is impossible since $|a - \mathrm{b}| = 2\varepsilon$.

**118.** (a) We show that any segment with center $a$ is a trough. Suppose that the length of the segment is equal to $2\varepsilon$. We need to show that infinitely many terms of the sequence lie inside this segment, that is, that the inequality $|x_n - a| < \varepsilon$ is satisfied. Let us suppose that this is not so. Then only finitely many terms of the sequence lie inside our segment.

Let $k$ be the number of the last position in which one of these terms occurs. (If no term of the sequence lies on our segment, we take $k = 0$.) Then all terms after the $k$th lie outside the segment. But this contradicts the definition of a limit point. For each $k$ there exists $n > k$ such that $|x_n - a| < \varepsilon$ and thus $x_n$ lies within the segment.

(b) Suppose that the opposite is true, that is, that $a$ is not a limit point of the sequence. Then for some $\varepsilon$ and some $k$ there is no number $n > k$ for which $|x_n - a| < \varepsilon$. In other words, there is a segment having center $a$ for which only finitely many (not more than $k$) terms of the sequence lie inside. We thus have a contradiction to our original assumption that each segment having center $a$ is a trough for $\{x_n\}$ and thus contains infinitely many terms of the sequence.

**119.** This follows from the solutions of Problems 113a, 107a, and 118b. Try to present this solution without using the concepts of trap and trough.

**120.** (a) The sequence $x_n = (n + 1)/n$ has the number 1 as limit. Thus, (see Problem 119), the number 1 is a limit point of this sequence. The sequence does not have any other limit points (see Problems 114a and 118a).

(b) The points $+1$ and $-1$, clearly, are limit points of the sequence $x_n = (-1)^n$. For any other point $a$ it is possible to construct a segment having center $a$ which contains not a single term of the sequence. Therefore the sequence has no other limit points.

(c) Since the function $\sin x°$ has period 360°, each of the numbers 0, $\pm\sin 1°$, $\pm\sin 2°, \ldots, \pm\sin 89°$, $\pm 1$ occurs an infinite number of times in the sequence. Therefore all of these numbers are limit points. If the number $a$ does not coincide with any of the above-listed 181 numbers, we can construct a segment having center $a$ which does not contain a single term of the sequence. (For this it suffices to take the length of the segment less than the distance from the point $a$ to the nearest of the numbers just mentioned.) Therefore the sequence has no other limit points.

(d) It is convenient to represent this sequence as the union of the two sequences $y_n = 1/(2n - 1)$ and $z_n = 2n$. The first sequence has the single limit point 0; the second has no limit points. (Prove this assertion by yourself.) We now show that the point $a$ is a limit point for the original sequence if and only if it is a limit point for one of the two sequences $\{y_n\}$ and $\{z_n\}$. In fact, if we use the result of Problem 118, it suffices to prove the following almost obvious proposition: The segment $[a, b]$ is a trough for the sequence $\{x_n\}$ if and only if it is a trough for at least one of the sequences $\{y_n\}$ and $\{z_n\}$.

**Remark.** By the same argument we prove the general theorem: If the sequence $\{x_n\}$ is the union of a finite number of sequences $\{y_n\}, \{z_n\}, \ldots, \{t_n\}$, the set of limit points of the sequence $\{x_n\}$ is the union of the sets of limit points of the sequences $\{y_n\}, \{z_n\}, \ldots, \{t_n\}$.

(e) If the number $a$ satisfies the conditions $0 \leq a \leq 1$, then infinitely many terms of our sequence lie inside each segment having center $a$. For if the length of the segment is equal to $2\varepsilon$, for any $n > 1/\varepsilon$ there is at least one term among the terms $1/n$, $2/n$, $\ldots, n - (1/n)$ inside our segment. If $a < 0$ or $a > 1$, it is possible to construct a segment having center $a$ which contains no term of the sequence.

Therefore the limit points of our sequence are precisely the points of the segment $[0, 1]$.

**122.** (a) Suppose that the sequence $\{x_n\}$ has limit $x_0$. We take a segment $[a, b]$ having center $x_0$. Since this segment is a trap for the sequence $\{x_n\}$ (see Problem 113a), there are only finitely many terms of the sequence that lie outside of this segment. Suppose that $x_k$ is the greatest (in absolute value) of these terms. We denote by $C$ the greatest of the numbers $|a|$, $|b|$, and $|x_k|$. Then for all terms of the sequence the inequality $|x_n| \leq C$ will be valid. For if $x_n$ lies inside the segment $[a, b]$, $|x_n|$ does not exceed the greater of the numbers $|a|$, $|b|$. If $x_n$ does not lie inside the segment $[a, b]$, $|x_n| \leq |x_k|$ (since $x_k$ is the term whose absolute value is the greatest of those of the terms not lying inside $[a, b]$). Thus, a fortiori, $|x_n| \leq C$.

(b) The sequence introduced in Problem 108b is bounded since $|x_n| \leq 2$. We show that it has no limit. Indeed, it is easy to verify that the segments $[0, \frac{1}{4}]$ and $[1, 1\frac{1}{4}]$ are troughs for this sequence. The distance between these troughs is equal to $\frac{3}{4}$.

It follows (see the solution to Problem 110a) that for this sequence there exists no trap of length less than $\frac{3}{4}$. But if the sequence $\{x_n\}$ had limit $a$, each segment with center $a$ would be a trap, and thus there exist traps of arbitrarily small length. This contradiction shows that the sequence $\{x_n\}$ does not have a limit.

It is, of course, possible to obtain this result without using the terms "trap" and "trough" (compare the solutions of Problems 116h and 116j). Do this yourself.

**123.** (a) Let $N$ be an arbitrary positive number. For all $n > N$ the inequality $|x_n| > N$ is valid. Therefore the sequence $\{x_n\}$ converges to infinity.

(b) The solution is the same as in case (a).

(c) The sequence is not bounded, since for any positive number $C$ there exists an integer $n$ for which $|x_n| > C$ (for $n$ it suffices to take any even number greater than $C$). But this sequence does not converge to infinity since the inequality $|x_n| > 1$ fails to hold for the odd terms of the sequence.

(d) Let $N$ be an arbitrary positive number. For all $n > N^2$ the inequality $x_n > N$ is satisfied. Consequently, $x_n \to \infty$.

(e) We prove that the sequence $\{x_n\}$ is bounded. More exactly, we show that $|x_n| \leq 5$. Since $x_n > 0$, $|x_n| = x_n$, and so we must show that $x_n \leq 5$. But this is so by the following relations:

$$\begin{aligned} 5 - x_n &= 5 - \frac{100n}{100 + n^2} \\ &= \frac{5}{100 + n^2}(100 + n^2 - 20n) \\ &= \frac{5(10 - n)^2}{100 + n^2} \geq 0. \end{aligned}$$

**125.** We shall analyze only two examples in detail.

(a) We show that any sequence $\{x_n\}$ satisfies Condition 1. In fact, we need to show that there exists a positive number $\varepsilon$, a number $k$, and a number $n > k$ such that $|x_n - a| < \varepsilon$. For $\varepsilon$ we take the number

$|x_1 - a| + 1$ and set $k = 0, n = 1$. Then the necessary condition is satisfied.

(b) We show that Condition 6 means that the sequence $\{x_n\}$ does not have $a$ as a limit. For the assertion "$a$ is the limit of the sequence $\{x_n\}$" implies that for each $\varepsilon > 0$ there exists $k$ such that for all $n \geq k$ the inequality $|x_n - a| < \varepsilon$ is valid. The negation of this assertion reads as follows: "It is not the case that for any $\varepsilon > 0$ there exists $k$ such that for all $n > k$ the inequality $|x_n - a| < \varepsilon$ is valid."

We can rephrase this statement thus: "For some $\varepsilon > 0$ there does not exist a $k$ such that for all $n > k$ the inequality $|x_n - a| < \varepsilon$ is valid." This in turn can be formulated as follows: "There exists $\varepsilon > 0$ such that for any $k$ the inequality $|x_n - a| < \varepsilon$ is not valid for every $n \geq k$." Finally, we can rephrase this as "There exists $\varepsilon > 0$ such that for each $k$ there exists $n > k$ for which $|x_n - a| \geq \varepsilon$."

We have thus obtained Condition 6. Note that this condition can be deduced from the definition of limit using the following rule: (1) Substitute the phrase "for each" for "there exists . . . such that"; (2) substitute the phrase "there exists . . . such that" for "for each"; (3) Change the inequality $|x_n - a| < \varepsilon$ to the reverse inequality $|x_n - a| \geq \varepsilon$. This rule is applicable to all of the remaining conditions. Verify that for each of these conditions it gives the formulation of the negation of this condition.

**126.** If the first sign is "+," all of the terms of the sequence are equal to $a$. Therefore all of the remaining signs are uniquely determined: The sequence has the number $a$ as limit and limit point, is bounded, and does not converge to infinity.

Consider now choices beginning with "$-$ +"; these signs indicate that the sequence converges to $a$. Then all of the remaining signs are determined: The sequence has limit point $a$ (Problem 119), is bounded (Problem 122a), and does not converge to infinity.

Consider now choices ending in "+." Here the sequence converges to infinity and this determines all of the remaining terms: The sequence is not identically equal to $a$, does not converge to $a$, does not have $a$ as limit point, and is not bounded.

The remaining choices have a "$-$" in the first, second, and fifth places. The other signs, it turns out, can be chosen arbitrarily. Examples are given in "Answers and Hints."

**127.** In "Answers and Hints" examples of sequences having limits and having either a greatest or a least term or both are given. It suffices to show, therefore, that there exists no sequence which has a limit but which has neither a greatest nor a least term.

Suppose that the sequence $\{x_n\}$ does not have a greatest term. We shall show that we can then extract from the sequence $\{x_n\}$ a monotonically increasing subsequence. For let us consider the first term of the sequence. Since $x_1$ is not the greatest term, there exists a term $x_{n_1} > x_1$. This term $x_{n_1}$ cannot be greater than all of the terms of the sequence which follow it (since otherwise the maximum of the first $n_1$ terms of the sequence would exceed each term of the sequence). Therefore there exists an integer $n_2 > n_1$ such that $x_{n_2} > x_{n_1}$. The term $x_{n_2}$ again cannot exceed each of the terms following it, and so on. Thus we can get an infinite increasing subsequence

$$x_1 < x_{n_1} < x_{n_2} < \ldots < x_{n_k} < \ldots.$$

Analogously, if the sequence $x_n$ does not have a least term, there exists an infinite, monotonically decreasing subsequence

$$x_1 > x_{m_1} > x_{m_2} > \ldots > x_{m_k} > \ldots.$$

Denote by $\varepsilon$ the difference between $x_{n_1}$ and $x_{m_1}$. Then for all $k > 1$ the inequality $x_{n_k} - x_{m_k} > x_{n_1} - x_{m_1} = \varepsilon$ is valid. From this it follows (see the solution to Problems 116h or 123b) that the sequence $\{x_n\}$ does not have a limit.

**128.** If the sequence $\{x_n\}$ or a subsequence of this sequence fails to have a greatest term, we can, as shown in the solution of Problem 127, find an increasing subsequence. Consider now the case where each subsequence has a greatest term. Let $x_{n_1}$ be the greatest term of the sequence, $x_{n_2}$ the greatest term of the sequence after $x_{n_1}$, $x_{n_3}$ the greatest term after $x_{n_2}$, and so on. Clearly, the inequalities

$$x_{n_1} \geq x_{n_2} \geq \ldots \geq x_{n_k} \geq \ldots$$

are valid.

Thus, in this case as well, it is possible to choose a monotone subsequence of the sequence $\{x_n\}$.

**129.** Consider the sequence 1.4, 1.41, 1.414, . . . . The term $x_n$ of this sequence is the value of $\sqrt{2}$ to $n$ decimal places.

From the very definition of the sequence $\{x_n\}$ it follows that $\lim_{n \to \infty} x_n = \sqrt{2}$ (for $n > \log(1/\varepsilon)$ the inequality $|x_n - \sqrt{2}| < \varepsilon$ is satisfied). Since $\sqrt{2}$ is irrational and since the sequence $\{x_n\}$ can have at most one limit, no rational number can be the limit of this sequence.

**Remark.** In this proof we have made use of the fact that there exists a real number whose square is equal to 2 and which can be written as an infinite decimal fraction. It is possible to change our argument so that only rational numbers need be used. For this we need to define $x_n$ as the greatest number that can be written as a finite decimal fraction having $n$ digits after the decimal point and whose square is less than 2. We must then prove that if $x_n$ converges to a number $a$, $a^2 = 2$.

The main idea of this proof is the following:

The number $x_n$ has the property that $x_n^2$ is less than 2, but $(x_n + 1/10^n)^2$ is greater than 2. Therefore for large $n$ the number $x_n^2$ is very close to 2. However, for sufficiently large $n$ the number $x_n$ is close to $a$, and thus the numbers $x_n^2$ and $a^2$ are nearly equal. From this we can conclude that the difference between $a^2$ and 2 is arbitrarily small. But this difference is a constant independent of $n$. Therefore $a^2 = 2$. It remains to prove only that there exists no rational number $a$ whose square is equal to 2. Try to formulate a rigorous proof based on the ideas developed here.

**130.** Suppose that we are given a bounded sequence. We choose a monotone subsequence of it as in the solution to Problem 128. This subsequence is bounded and by the Bolzano-Weierstrass axiom has a limit. We show that this limit is a limit point of the original sequence.

In fact, each segment having this point as center is a trap for the subsequence that we have chosen. Therefore infinitely many of the terms of the original sequence lie on this segment, and so this segment is a trough for the original sequence. Our assertion now follows from the theorem of 118b.

**131.** (a) The sequence $x_n = 1 + \frac{1}{4} + \ldots + 1/n^2$ is clearly increasing. We show that it is bounded. Indeed, since $1/n^2 < 1/n(n-1)$,

$$x_n < 1 + \frac{1}{1\cdot 2} + \frac{1}{2\cdot 3} + \ldots + \frac{1}{n(n-1)}$$

$$= 1 + (1 - \tfrac{1}{2}) + (\tfrac{1}{2} - \tfrac{1}{3}) + \ldots + \left(\frac{1}{n-1} - \frac{1}{n}\right)$$

$$= 2 - \tfrac{1}{n} > 2.$$

Thus, $|x_n| < 2$ for all $n$. Therefore, from the Bolzano-Weierstrass axiom it follows that $\lim_{n\to\infty} x_n$ exists.

Using the methods of higher mathematics, one can establish that this limit is equal to $\pi^2/6$. For an elementary proof of this, see the book by A. M. Yaglom and I. M. Yaglom, *Challenging Mathematical Problems with Elementary Solutions.*[1]

(b) It is clear that the subsequence $x_{2n}$ of even-numbered terms is increasing, since

$$x_{2n} = (1 - \tfrac{1}{3}) + (\tfrac{1}{5} - \tfrac{1}{7}) + \ldots + \left(\frac{1}{4n-3} - \frac{1}{4n-1}\right)$$

$$> x_{2n-2},$$

and bounded, since

$$x_{2n} = 1 - (\tfrac{1}{3} - \tfrac{1}{5}) - (\tfrac{1}{7} - \tfrac{1}{9}) - \ldots -$$

$$- \left(\frac{1}{4n-5} - \frac{1}{4n-3}\right) - \frac{1}{4n-1} < 1.$$

[1]Revised edition edited by B. Gordon, Holden-Day, Inc., San Francisco, 1969.

Therefore $\lim_{n \to \infty} x_{2n}$ exists. We denote this limit by $a$ and show that the number $a$ is the limit of the full sequence $\{x_n\}$. For suppose that we are given a positive number $\varepsilon$. Then there exists a number $k$ such that for $2n > k$ the inequality $|x_{2n} - a| < \varepsilon/2$ is satisfied. Moreover, for $2n + 1 > 1/\varepsilon$ the inequality $|x_{2n+1} - x_{2n+2}| < \varepsilon/2$ is satisfied. Therefore, if the number $n$ is greater than the maximum of $k$ and $1/\varepsilon$, $|x_n - a| < \varepsilon$.

It turns out that $a = \pi/4$ (for a proof, see, for example, the above-mentioned book by Yaglom and Yaglom).

**132.** (a) We are required to prove that for each positive number $\varepsilon$ there exists $k$ such that for all $n > k$, $|x_n + y_n - a - b| < \varepsilon$. Since by assumption $\lim_{n \to \infty} x_n = a$, there exists an integer $k_1$ such that $|x_n - a| < \varepsilon/2$ for $n > k_1$.

Similarly, from the fact that $\lim_{n \to \infty} y_n = b$, we know that there exists a number $k_2$ such that $|y_n - b| < \varepsilon/2$ for $n > k_2$. Let $k$ be the greater of the numbers $k_1$ and $k_2$. Then for $n > k$ we get

$$|x_n + y_n - a - b| \leq |x_n - a| + |y_n - b| < \varepsilon.$$

(b) The proof is the same as in (a).

(c) We use the identity

$$\begin{aligned} x_n y_n - ab &= x_n y_n - b x_n + b x_n - ab \\ &= x_n(y_n - b) + b(x_n - a). \end{aligned}$$

Since the sequence $x_n$ has a limit, it is bounded (see Problem 122a), that is, there exists $C$ such that $|x_n| \leq C$ for all $n$. We denote by $M$ the greater of the numbers $C$ and $|b|$. Suppose now that we are given an arbitrary positive number $\varepsilon$. Since $\lim_{n \to \infty} x_n = a$, there exists a number $k_1$ such that for $n > k_1$ the inequality $|x_n - a| < \varepsilon/2M$ is satisfied. In the same way there exists a number $k_2$ such that the inequality $|y_n - b| < \varepsilon/2M$ is satisfied for $n > k_2$. Let $k$ denote the greater of the numbers $k_1$ and $k_2$. Then for $n > k$ we have

$$\begin{aligned}|x_n y_n - ab| &= |x_n(y_n - b) + b(x_n - a)| \\ &\leq |x_n|\,|y_n - b| + |b|\,|x_n - a| \\ &\leq M\frac{\varepsilon}{2M} + M\frac{\varepsilon}{2M} = \varepsilon.\end{aligned}$$

(d) We use the identity

$$\frac{x_n}{y_n} - \frac{a}{b} = \frac{x_n - a}{y_n} + \frac{a(b - y_n)}{by_n}.$$

Suppose first that we have been able to prove the boundedness of the sequence $\{1/y_n\}$. Then we know that there exists a number $M$ such that $|1/y_n| \leq M$ for all $n$. Then

$$\begin{aligned}\left|\frac{x_n}{y_n} - \frac{a}{b}\right| &\leq \left|\frac{x_n - a}{y_n}\right| + \left|\frac{a(b - y_n)}{by_n}\right| \\ &\leq M\,(x_n - a) + M\left|\frac{a}{b}\right|\left|y_n - b\right|.\end{aligned}$$

Suppose now that we are given an arbitrary positive number $\varepsilon$. We can find a number $k_1$ such that $|x_n - a| < \varepsilon/2M$ for $n > k_1$ and a number $k_2$ such that $|y_n - b| < \varepsilon\,|b|/(2\,|a|\,M)$ for $n > k_2$. Denote by $k$ the greater of the numbers $k_1$ and $k_2$. Then for $n > k$ we get

$$\left|\frac{x_n}{y_n} - \frac{a}{b}\right| \leq M\frac{\varepsilon}{2M} + M\left|\frac{a}{b}\right|\frac{\varepsilon\,|b|}{2\,|a|\,M} = \varepsilon.$$

We now prove that if $\lim_{n\to\infty} y_n = b$, $b \neq 0$, $y_n \neq 0$, the sequence $\{1/y_n\}$ is bounded. Suppose for definiteness that $b > 0$. Then there exists a number $k$ such that the inequality $|y_n - b| < b/2$ is satisfied for $n > k$. This inequality then implies that for $n > k$ all of the terms of the sequence $\{y_n\}$ lie inside the segment $[b/2, 3b/2]$. Denote by $M$ the maximum of the numbers $2/b$, $1/|y_1|$, $1/|y_2|, \ldots, 1/|y_k|$ (we assume here that the number $k$ has been chosen to be an integer). Clearly, the inequality $|1/y_n| \leq M$ is satisfied for all $n$. Analyze the case $b < 0$ by yourself.

**134.** Using the results of Problem 132, we can write

(a) $$\lim_{n\to\infty} \frac{2n+1}{3n-5} = \lim_{n\to\infty} \frac{2+\frac{1}{n}}{3-\frac{5}{n}} = \frac{\lim\limits_{n\to\infty}\left(2+\frac{1}{n}\right)}{\lim\limits_{n\to\infty}\left(3-\frac{5}{n}\right)} = \frac{2}{3};$$

(b) $$\lim_{n\to\infty} \frac{10n}{n^2+1} = \lim_{n\to\infty} \frac{\frac{10}{n}}{1+\frac{1}{n^2}} = \frac{\lim\limits_{n\to\infty}\left(\frac{10}{n}\right)}{\lim\limits_{n\to\infty}\left(1+\frac{1}{n^2}\right)} = 0;$$

(c) $$\lim_{n\to\infty} \frac{n(n+2)}{(n+1)(n+3)} = \lim_{n\to\infty} \frac{1+\frac{2}{n}}{\left(1+\frac{1}{n}\right)\left(1+\frac{3}{n}\right)}$$

$$= \frac{\lim\limits_{n\to\infty}\left(1+\frac{2}{n}\right)}{\lim\limits_{n\to\infty}\left(1+\frac{1}{n}\right)\lim\limits_{n\to\infty}\left(1+\frac{3}{n}\right)} = 1;$$

(d) $$\lim_{n\to\infty} \frac{2^n-1}{2^n+1} = \lim_{n\to\infty} \frac{1-\frac{1}{2^n}}{1+\frac{1}{2^n}} = \frac{\lim\limits_{n\to\infty}\left(1-\frac{1}{2^n}\right)}{\lim\limits_{n\to\infty}\left(1+\frac{1}{2^n}\right)} = 1.$$

(e) As shown in Chapter 1, the sum $1^k + 2^k + \ldots + n^k$ is a polynomial of degree $k+1$ in $n$ with leading coefficient $[1/(k+1)]n^{k+1}$ (see Problem 23). Therefore,

$$\lim_{n\to\infty} x_n = \lim_{n\to\infty} \frac{\frac{1}{k+1}n^{k+1} + a_1 n^k + \ldots + a_{k+1}}{n^{k+1}}$$

$$= \lim_{n\to\infty}\left(\frac{1}{k+1} + \frac{a_1}{n} + \frac{a_2}{n^2} + \ldots + \frac{a_{k+1}}{n^{k+1}}\right)$$

$$= \frac{1}{k+1}.$$

**135.** We have

$$\sqrt{n+1} - \sqrt{n-1} = \frac{2}{\sqrt{n+1} \quad \sqrt{n+1}}$$

(see "Answers and Hints"). Therefore,

$$|x_n| = \frac{2}{\sqrt{n+1} + \sqrt{n-1}} \leq \frac{1}{\sqrt{n-1}}.$$

It follows that for $n > 1 + 1/\varepsilon^2$ the inequality $|x_n| < \varepsilon$ is satisfied.

**136.** We use the fact that

$$2^n = (1+1)^n = 1 + n + \frac{n|n-1|}{2} + \dots$$

$$> n + \frac{n(n-1)}{2} = \frac{n^2}{2}.$$

Therefore, $|x_n| = n/2^n < 2/n$. For $n > 2/\varepsilon$ the inequality $|x_n| < \varepsilon$ is satisfied.

**137.** We use the fact that

$$a^n = [1 + (a-1)]^n = 1 + n(a-1) +$$

$$+ \frac{n(n-1)}{2}(a-1)^2 + \dots > \frac{n(n-1)}{2}(a-1)^2$$

Therefore,

$$|x_n| = \frac{n}{a^n} < \frac{n}{\frac{n(n-1)}{2}(a-1)^2} = \frac{2}{(n-1)(a-1)^2}.$$

For any positive $\varepsilon$ we will have $|x_n| < \varepsilon$ whenever $n > 2/[(a-1)^2\varepsilon] + 1$.

**138.** Suppose that the number $n$ lies between $2^{m-1}$ and $2^m$; then $\log_2 n$ lies between $m - 1$ and $m$. Consequently, the expression $\log_2 n/n$ does not exceed the quantity $m/2^{m-1}$. But we know (see Problem 136) that for any $\varepsilon > 0$ there exists $k$ such that for $m > k$ the inequality $m/2^m < \varepsilon/2$, or $m/2^{m-1} < \varepsilon$, is satisfied. We show that for $n > 2^k$ the inequality $\log_2 n/n < \varepsilon$ is satisfied. For if $n > 2^k$, $n$ lies between $2^{m-1}$ and $2^m$, where $m > k$. Therefore,

$$\frac{\log_2 n}{n} < \frac{m}{2^{m-1}} < \varepsilon$$

**139.** We show that for each positive $\varepsilon$ there exists a number $k$ such that for $n > k$ the inequality $|\sqrt[n]{n} - 1| < \varepsilon$ is satisfied. Since $\sqrt[n]{n} \geq 1$,

$$|\sqrt[n]{n} - 1| = \sqrt[n]{n} - 1.$$

Let us see for which $n$ the inequality $\sqrt[n]{n} - 1 \geq \varepsilon$ is satisfied, in other words, for which the inequality $\sqrt[n]{n} - 1 \geq \varepsilon$, or $\sqrt[n]{n} \geq 1 + \varepsilon$, or $n \geq (1 + \varepsilon)^n$, is satisfied. Since

$$(1 + \varepsilon)^n = 1 + n\varepsilon + \frac{n(n-1)}{2}\varepsilon^2 + \ldots > \frac{n(n-1)}{2}\varepsilon^2,$$

the inequality $n \geq (1 + \varepsilon)^n$ implies that $n > n(n-1)/2\,\varepsilon^2$; whence $n < 1 + 2/\varepsilon^2$. Therefore for all $n > 1 + (2/\varepsilon^2)$ the inequality $\sqrt[n]{n} - 1 \geq \varepsilon$ is false, and consequently, the inequality $\sqrt[n]{n} - 1 < \varepsilon$ is valid.

**140.** We shall use the symbols adopted in the hints for this problem. After the first step, the snail stops at the point $(a + 1, b)$; after the second, at the point $(a + 1, b + 1)$, and so on. After the $2n - th$ step it will be at the point $(a + n, b + n)$, and after the $2n + 1 - th$ step it will be at the point $(a + n + 1, b + n)$. Therefore,

$$k_{2n} = \frac{b + n}{a + n}, \quad k_{2n+1} = \frac{b + n}{a + n + 1}.$$

It remains only to find the limit of the sequence $\{k_n\}$. We denote $k_{2n}$ by $k_n'$ and $k_{2n+1}$ by $k_n''$. It is easy to verify (see the solution to Problem 134) that the sequences $\{k_n'\}$ and $\{k_n''\}$ converge to 1. Therefore, for each $\varepsilon > 0$ there exists a number $l_1$ such that for $n > l_1$ the inequality $|k_n' - 1| < \varepsilon$ will be satisfied. In the same way there exists a number $l_2$ such that for $n > l_2$ the inequality $|k_n'' - 1| < \varepsilon$ will be satisfied. Denote by $l$ the greater of the number $2l_1$ and $2l_2 + 1$. Then for $n > l$ the inequality $|k_n - 1| < \varepsilon$ will be satisfied.

**141.** We keep the notations used in the hint and solution for Problem 140.

(a) In this case

$$k_n' = k_{2n} = \frac{b + 2n}{a + n}, \quad k_n'' = k_{2n+1} = \frac{b + 2n}{a + n + 1}.$$

We obtain $\lim_{n\to\infty} k_n' = \lim_{n\to\infty} k_n'' = 2$. From this, as before, we deduce that $\lim_{n\to\infty} k_n$ exists and is equal to 2.

(b) In this case

$$k_n' = k_{2n} = \frac{b + 2 + 4 + \ldots + 2n}{a + 1 + 3 + 5 + \ldots + (2n - 1)}$$

$$= \frac{b + n^2 + n}{a + n^2};$$

$$k_n'' = k_{2n+1} = \frac{b + 2 + 4 + \ldots + 2n}{a + 1 + 3 + \ldots + (2n + 1)}$$

$$= \frac{b + n^2 + n}{a + (n + 1)^2}.$$

From this, $\lim_{n\to\infty} k_n' = \lim_{n\to\infty} k_n'' = 1$ and, as formerly, we prove that $\lim_{n\to\infty} k_n$ exists and is equal to 1.

(c) In this case

$$k_n' = k_{2n} = \frac{b + 2 + 8 + \ldots + 2^{2n-1}}{a + 1 + 4 + \ldots + 2^{2n-2}}$$

$$= \frac{b + \frac{2}{3}(4^n - 1)}{a + \frac{1}{3}(4^n - 1)};$$

$$k_n'' = k_{2n+1} = \frac{b + 2 + 8 + \ldots + 2^{2n-1}}{a + 1 + 4 + \ldots + 2^{2n}}$$

$$= \frac{b + \frac{2}{3}(4^n - 1)}{a + \frac{1}{3}(4^{n+1} - 1)}.$$

From this it is clear that $\lim_{n\to\infty} k'_n = 2$, $\lim_{n\to\infty} k''_n = \frac{1}{2}$. We show that the sequence $\{k_n\}$ does not have a limit in this case. For in fact, if a sequence has a limit $a$, each subsequence has the same limit $a$. Prove this by yourself, using, for example, Problems 113a and 113b.

But in our case the subsequences $\{k_{2n}\}$ and $\{k_{2n+1}\}$ have different limits.

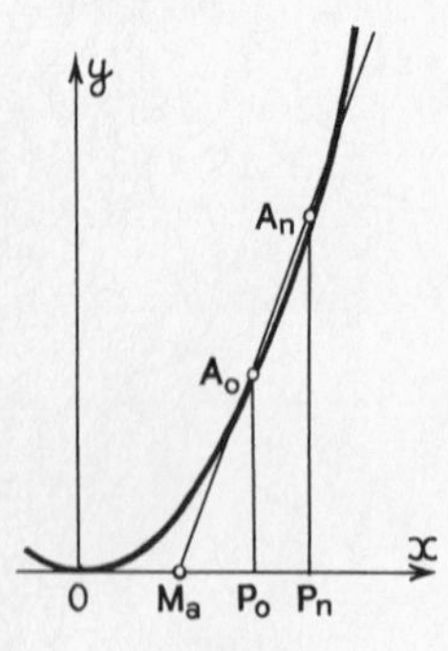

Fig. 38

**142.** From the similarity of triangles $A_0M_0P_0$ and $A_nM_nP_n$ (Fig. 38) we get

$$M_nP_0 : M_nP_n = P_0A_0 : P_nA_n.$$

Denote the abscissa of the point $M_n$ by $x_n$. Then the equality just obtained can be rewritten:

$$(a - x_n) : \left(a + \frac{1}{n} - x_n\right) = a^2 : \left(a + \frac{1}{n}\right)^2.$$

From this we get the expression for $x_n$:

$$x_n = \frac{a(an + 1)}{2an + 1} = \frac{a\left(a + \frac{1}{n}\right)}{2a + \frac{1}{n}}.$$

Therefore,

$$\lim_{n\to\infty} x_n = \frac{\lim_{n\to\infty} a\left(a + \frac{1}{n}\right)}{\lim_{n\to\infty}\left(2a + \frac{1}{n}\right)} = \frac{a^2}{2a} = \frac{a}{2}.$$

The answer we have obtained can be formulated geometrically. The tangent to any point of a parabola bisects the segment joining the vertex of the parabola with the projection of the point of tangency onto the $x$-axis.

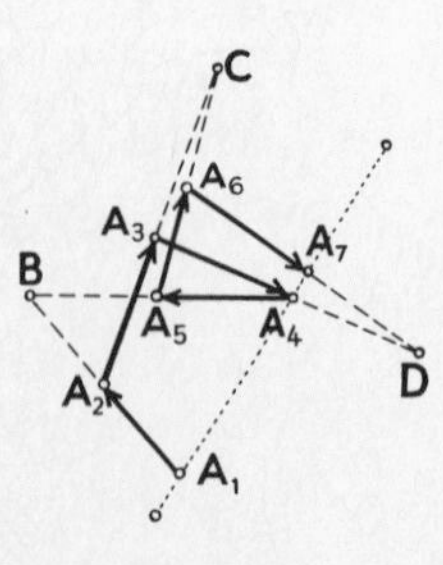

Fig. 39

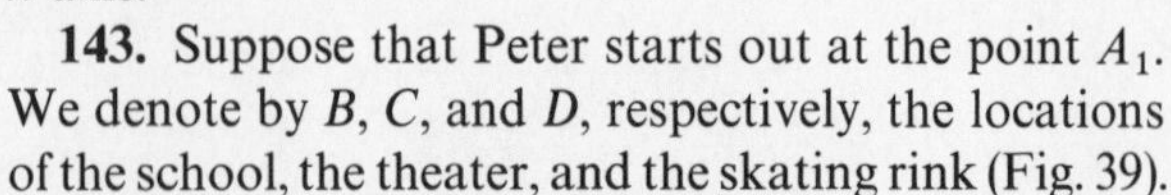

**143.** Suppose that Peter starts out at the point $A_1$. We denote by $B$, $C$, and $D$, respectively, the locations of the school, the theater, and the skating rink (Fig. 39).

Peter's path is represented by the broken line $A_1A_2A_3\ldots$, in which the point $A_2$ is the midpoint of the segment $A_1B$, the point $A_3$ is the midpoint of the segment $A_2C$, the point $A_4$ is the midpoint of the segment $A_3D$, and so on. Let us prove that the points $A_1$, $A_4, A_7, \ldots, A_{3n+1}$ lie on a straight line. For in fact, the segment $A_2A_5$ is the line through the midpoints of two sides of the triangle $A_1BA_4$ and therefore is parallel to the third side $A_1A_4$ and is half of its length. In the same way, considering the triangle $A_2CA_5$, we deduce that the segment $A_3A_6$ is parallel to $A_2A_5$ and has half its length. Finally, from triangle $A_3DA_6$ we find that the segment $A_4A_7$ is parallel to $A_3A_6$ and is half as long. Comparing the results, we see that segments $A_1A_4$ and $A_4A_7$ are continuations of one another and that the length of $A_4A_7$ is one-eighth that of $A_1A_4$.

In the same way we can establish that the segment $A_7A_{10}$ is the continuation of the segment $A_4A_7$, the segment $A_{10}A_{13}$ is the continuation of segment $A_7A_{10}$, and so on. Thus, the points $A_1, A_4, A_7, \ldots, A_{3n+1}, \ldots$ lie on a straight line. We have also proved that the distances between these points form a geometric progression with ratio $\frac{1}{8}$. It is now easy to show that the sequence $A_1, A_4, A_7 \ldots, A_{3n+1}, \ldots$ has a limit. For suppose that we take the point $A_1$ as the origin on the line passing through the points of our sequence and the segment $A_1A_4$ as the unit of length. Then the coordinate of the point $A_{3n+1}$ will be equal to

$$x_n = 1 + \frac{1}{8} + \frac{1}{8^2} + \ldots + \frac{1}{8^{n-1}} = \frac{1 - \left(\frac{1}{8}\right)^n}{1 - \frac{1}{8}}.$$

Therefore, $\lim_{n\to\infty} x_n$ exists and is equal to $1\frac{1}{7}$ (see Problem 132).

We have thus proved that the sequence of points $A_1$, $A_4, A_7, \ldots, A_{3n+1}$ converges to a point $M$. In the same way we can prove that the sequence $A_2, A_5, A_8, \ldots,$ $A_{3n+2}, \ldots$ converges to a point $N$ and that the sequence $A_3, A_6, A_9, \ldots, A_{3n}$ converges to a point $P$. Thus,

Peter's path begins to approximate the triangle $MNP$ rather soon. Prove by yourself that the positions of the points $M$, $N$, and $P$ depend only on the positions of the points $B$, $C$, and $D$ and not on the point $A_1$ from which the motion starts.

**144. First Method.** We draw a system of coordinates on the line $M_1M_2$, taking the point $M_1$ as origin and the segment $M_1M_2$ as the unit of measure. Then the coordinate $x_n$ of the point $M_n$ is connected to the coordinates of the preceding points by the relation

$$x_n = \frac{x_{n-1} + x_{n-2}}{2}.$$

(verify this equation by yourself). We prove that the sequence $\{x_n\}$ converges to $\frac{2}{3}$.

For this we show by induction that

$$x_n = \tfrac{2}{3}[1 + 2(-\tfrac{1}{2})^n].$$

In fact, setting $n = 1$, we get $x_1 = \frac{2}{3}[1 + 2(-\frac{1}{2})] = 0$, and for $n = 2$, $x_2 = \frac{2}{3}[1 + 2(-\frac{1}{2})^2] = 1$.

Suppose now that the equality $x_n = \frac{2}{3}[1 + 2(-\frac{1}{2})^n]$ is proved for all $n \leq k$, and let us check its validity for $n = k + 1$.

Substituting the expression for $x_k$ and $x_{k-1}$ in the formula $x_{k+1} = (x_k + x_{k-1})/2$, we obtain

$$\begin{aligned} x_{k+1} &= \frac{\frac{2}{3}[1 + 2(-\frac{1}{2})^k] + \frac{2}{3}[1 + 2(-\frac{1}{2})^{k-1}]}{2} \\ &= \tfrac{2}{3}[1 + (-\tfrac{1}{2})^k + (-\tfrac{1}{2})^{k-1}] \\ &= \tfrac{2}{3}[1 + (-\tfrac{1}{2})^k(1 - 2)] = \tfrac{2}{3}[1 + 2(-\tfrac{1}{2})^{k+1}]. \end{aligned}$$

**Second Method.** Consider the point $N$ that divides the segment $M_1M_2$ in the ratio 2:1. Prove by induction that this point divides each of the segments $M_2M_5$, $M_3M_4, \ldots, M_nM_{n+1}, \ldots$ in the ratio 2:1. From this it follows directly that the length of the segments $M_nN$ is $2^{1-n}$ times the length of segment $M_1N$. But this means that the sequence $M_1, M_2, \ldots, M_n, \ldots$ has the point $N$ as its limit.

**145.** (a) By the formula for the sum of a geometric progression $S_n = (1 - a^{n+1})/(1 - a)$. If $|a| < 1$,

$$\lim_{n\to\infty} S_n = \frac{\lim_{n\to\infty}(1 - a^{n+1})}{1 - a} = \frac{1}{1 - a}.$$

If $|a| > 1$, $S_n \to \infty$ as $n \to \infty$. Finally, if $|a| = 1$, we get either the sequence $S_n = n$ (for $a = 1$) or $S_n = [1 + (-1)^n]/2$ (for $a = -1$). Both of these sequences fail to have limits. Thus, for $|a| \geq 1$ the series diverges.

(b) We represent $S_n$ in the form $(a + a^2 + \ldots + a^n) + (a^2 + a^3 + \ldots + a^n) + \ldots + (a^{n-1} + a^n) + a^n$. Applying the formula for the sum of a geometric progression in each of the expressions in parentheses, we get

$$S_n = \frac{a - a^{n+1}}{1 - a} + \frac{a^2 - a^{n+1}}{1 - a} + \ldots + \frac{a^n - a^{n+1}}{1 - a}$$

$$= \frac{a + a^2 + \ldots + a^n - na^{n+1}}{1 - a}$$

$$= \frac{\dfrac{a - a^{n+1}}{1 - a} - na^{n+1}}{1 - a}$$

$$= \frac{a - (n + 1 - na)\,a^{n+1}}{(1 - a)^2}.$$

From this formula it is easily deduced (see Problem 137) that for $|a| < 1$, $S_n \to a/(1 - a)^2$, and for $|a| \geq 1$ the series diverges.

(c) Since $1/n(n + 1) = 1/n - 1/(n + 1)$,

$$S_n = \left(1 - \frac{1}{2}\right) + \left(\frac{1}{2} - \frac{1}{3}\right) + \ldots + \left(\frac{1}{n} - \frac{1}{n + 1}\right)$$

$$= 1 - \frac{1}{n + 1};\ S = 1.$$

(d) Since

$$\frac{1}{n(n + 1)(n + 2)} = \frac{1}{2}\left[\frac{1}{n(n + 1)} - \frac{1}{(n + 1)(n + 2)}\right],$$

$$S_n = \frac{1}{2}\left[\left(\frac{1}{1\cdot 2} - \frac{1}{2\cdot 3}\right) + \left(\frac{1}{2\cdot 3} - \frac{1}{3\cdot 4}\right) + \ldots + \right.$$

$$\left. + \frac{1}{n(n+1)} - \frac{1}{(n+1)(n+2)}\right]$$

$$= \frac{1}{2}\left[\frac{1}{1\cdot 2} - \frac{1}{(n+1)(n+2)}\right]; \qquad S = \frac{1}{4}.$$

(e) We note that

$$\frac{1}{n+1} + \frac{1}{n+2} + \ldots + \frac{1}{2n} > \frac{1}{2}.$$

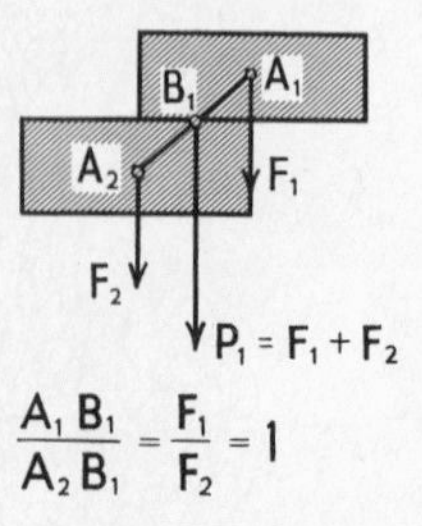


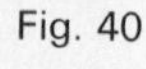
Fig. 40

For the left side is the sum of $n$ terms, the least of which is equal to $1/2n$. Furthermore,

$$S_2^n = 1 + \frac{1}{2} + \left(\frac{1}{3} + \frac{1}{4}\right) + \left(\frac{1}{5} + \frac{1}{6} + \frac{1}{7} + \frac{1}{8}\right) +$$

$$+ \ldots + \left(\frac{1}{2^{n-1}+1} + \ldots + \frac{1}{2^n}\right) > 1 + \frac{n}{2}$$

(as each of the expressions in parentheses is greater than $\frac{1}{2}$). It follows that $S_n \to \infty$ as $n \to \infty$. Thus, the series diverges.

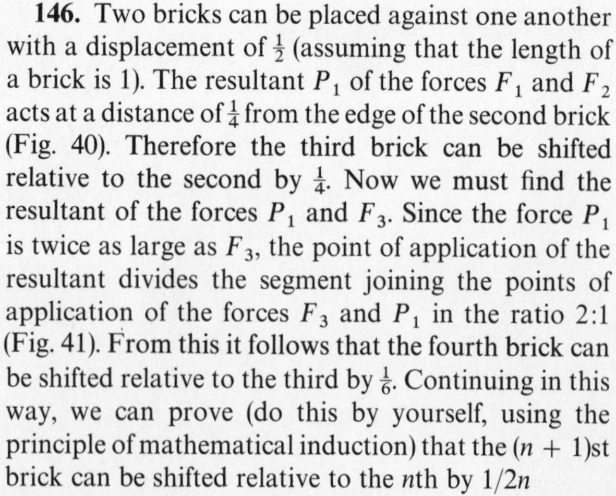

**146.** Two bricks can be placed against one another with a displacement of $\frac{1}{2}$ (assuming that the length of a brick is 1). The resultant $P_1$ of the forces $F_1$ and $F_2$ acts at a distance of $\frac{1}{4}$ from the edge of the second brick (Fig. 40). Therefore the third brick can be shifted relative to the second by $\frac{1}{4}$. Now we must find the resultant of the forces $P_1$ and $F_3$. Since the force $P_1$ is twice as large as $F_3$, the point of application of the resultant divides the segment joining the points of application of the forces $F_3$ and $P_1$ in the ratio 2:1 (Fig. 41). From this it follows that the fourth brick can be shifted relative to the third by $\frac{1}{6}$. Continuing in this way, we can prove (do this by yourself, using the principle of mathematical induction) that the $(n + 1)$st brick can be shifted relative to the $n$th by $1/2n$

In this way it is possible to construct a "roof" of $n$ bricks of length $\frac{1}{2} + \frac{1}{4} + \frac{1}{6} + \ldots + 1/2(n - 1)$. Since

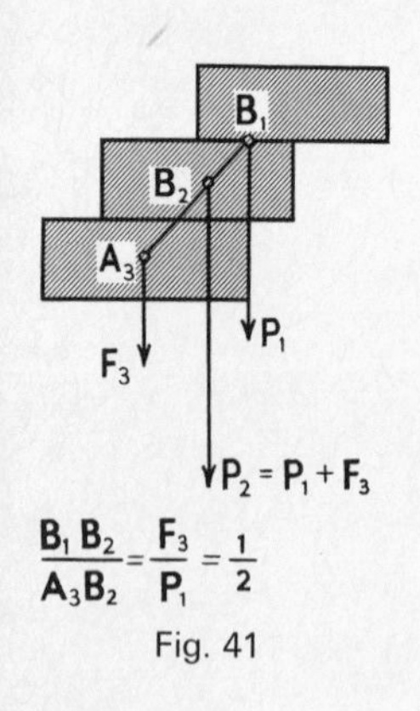


Fig. 41

the series $1 + \frac{1}{2} + \frac{1}{3} + \ldots + \frac{1}{n} + \ldots$ diverges (see Problem 145e), this "roof" can be made arbitrarily long.

**147.** The sequence $\{x_n\}$ clearly satisfied the relation $x_{n+1} = 2 + 1/x_n$. Suppose that the sequence $\{x_n\}$ has limit $a$. Then the left part of the equation converges to the number $a$, but the right side converges to the number $2 + 1/a$ (see Problem 132). We thus get the equality $a = 2 + 1/a$, whence $a = 1 \pm \sqrt{2}$. Since each $x_n$ is greater than 2, the number $1 - \sqrt{2}$ cannot be a limit of the sequence $\{x_n\}$.

Thus we have proved that if $\{x_n\}$ has a limit, this limit is equal to $1 + \sqrt{2}$.

We prove now that the desired limit actually exists. Denote by $y_n$ the difference between $x_n$ and $1 + \sqrt{2}$. We need to prove that $\lim_{n \to \infty} y_n = 0$. Substituting the expression for $x_n$ in terms of $y_n$ in the equality $x_{n+1} = 2 + 1/x_n$ we obtain

$$1 + \sqrt{2} + y_{n+1} = 2 + \frac{1}{1 + \sqrt{2} + y_n},$$

whence

$$y_{n+1} = \frac{2(1 + \sqrt{2} + y_n) + 1 - (1 + \sqrt{2})(1 + \sqrt{2} + y_n)}{1 + \sqrt{2} + y_n} = \frac{(1 - \sqrt{2})y_n}{1 + \sqrt{2} + y_n}.$$

From this formula we get the following estimate: $|y_{n+1}| < \frac{1}{2}|y_n|$. In fact, $|1 - \sqrt{2}| < \frac{1}{2}$, and $1 + \sqrt{2} + y_n > 1$ (this follows from the fact that $|y_n| < \frac{1}{2}$, as is easily established by induction). Therefore,

$$|y_{n+1}| = \frac{|1 - \sqrt{2}|\,|y_n|}{|1 + \sqrt{2} + y_n|} < \frac{1}{2}|y_n|.$$

From this we get directly that $|y_n| < y_1/2^{n-1} < 1/2^n$, and consequently the sequence $\{y_n\}$ converges to zero.

**Remark.** It is possible to prove that the sequence

$$n_1, \quad n_1 + \frac{1}{n_2}, \quad n_1 + \cfrac{1}{n_2 + \cfrac{1}{n_3}}, \quad n_1 + \cfrac{1}{n_2 + \cfrac{1}{n_3 + \cfrac{1}{n_4}}}, \ldots,$$

where $n_1, n_2, n_3, \ldots$ are arbitrary natural numbers, always has as limit some irrational number $a$. If the sequence $n_1$, $n_2$, $n_3, \ldots$ is periodic, the number $a$ has the form $r_1 + \sqrt{r_2}$, where $r_1$ and $r_2$ are rational numbers. The converse is also true: Each irrational number of the form $r_1 + \sqrt{r_2}$ can be written in the form of an infinite periodic continued fraction. For more on this subject see A. Ya. Khinchin's book, *Continued Fractions.*

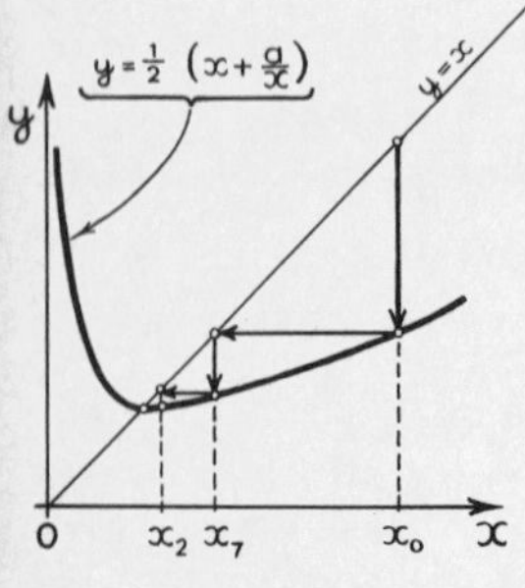

Fig. 42

**148.** First we prove that if $\{x_n\}$ has a limit, it is equal to $\pm\sqrt{a}$. For suppose that $\lim_{n\to\infty} x_n = b$. Then $\lim \frac{1}{2}[x_n + (a/x_n)] = \frac{1}{2}[b + (a/b)]$. We get the equality $b = \frac{1}{2}[b + (a/b)]$, whence $b^2 = a$, $b = \pm\sqrt{a}$.

It remains only to notice that if $x_0 > 0$, each term of the sequence is positive; if $x_0 < 0$, each term of the sequence is negative. Thus in the first case $\lim_{n\to\infty} x_n = \sqrt{a}$, and in the second case $\lim_{n\to\infty} x_n = -\sqrt{a}$.

There remains to be proved that the limit of the sequence $\{x_n\}$ actually exists. Suppose for definiteness that $x_0 > 0$ (Fig. 42). Denote by $y_n$ the difference between $x_n$ and $\sqrt{a}$ divided by $\sqrt{a}$. Substituting the expression $x_n = \sqrt{a}(1 + y_n)$ in the equality $x_{n+1} = \frac{1}{2}[x_n + (a/x_n)]$, we get

$$\sqrt{a}(1 + y_{n+1}) = \tfrac{1}{2}[\sqrt{a}(1 + y_n)] + \frac{a}{\sqrt{a}(1 + y_n)},$$

whence

$$y_{n+1} = \frac{y_n^2}{2(1 + y_n)}.$$

We need to prove that the sequence $\{y_n\}$ converges to zero. Note first that since

$$1 + y_0 = 1 + \frac{x_0 - \sqrt{a}}{\sqrt{a}} = \frac{x_0}{\sqrt{a}} > 0,$$

all of the terms $y_n$ for $n > 1$ are positive. Therefore

$$|y_{n+1}| = y_{n+1} = \frac{y_n^2}{2(1 + y_n)} < \frac{y_n}{2}.$$

From this it follows that $\lim_{n \to \infty} y_n = 0$.

Consider the concrete example $a = 10$, $x_0 = 3$. In this case $y_0 = (3 - \sqrt{10})/\sqrt{10}$. Since $(3.2)^2 = 10.24 > 10$, $\sqrt{10} < 3.2$. Therefore, $|y_0| = |(3 - \sqrt{10})/\sqrt{10}| < 0.2/3 = \frac{1}{15}$ and, therefore,

$$|y_1| = \frac{y_0^2}{2(1 + y_0)} < \frac{(\frac{1}{15})^2}{2(1 - \frac{1}{15})} < \frac{1}{400}.$$

Furthermore,

$$|y_2| = \frac{y_1^2}{2(1 + y_1)} < \frac{(\frac{1}{400})^2}{2} < \frac{1}{320{,}000}.$$

Therefore, $|x_2 - \sqrt{10}| = y_2\sqrt{10} < \sqrt{10}/320{,}000 < 0.00001$. Thus, to find $\sqrt{10}$ to an accuracy of 0.00001, it suffices to find the term $x_2$. We have

$$x_1 = \frac{1}{2}\left(x_0 + \frac{10}{x_0}\right) = \frac{1}{2}\left(3 + \frac{10}{3}\right)$$
$$= 3\tfrac{1}{6} = 3.166666\ldots,$$

$$x_2 = \frac{1}{2}\left(x_1 + \frac{10}{x_1}\right) = \frac{1}{2}\left(3\tfrac{1}{6} + \frac{10}{3\frac{1}{6}}\right)$$
$$3\tfrac{37}{228} = 3.162280\ldots.$$

In fact, $\sqrt{10} = 3.16227765\ldots$.

As you see, the value we have found really differs from the true value by less than 0.00001.

## Answers and Hints

### Chapter 1

**2.** The problem can be solved in two ways: by induction or by the formula

$$\frac{1}{n(n-1)} = \frac{1}{n} - \frac{1}{n+1}.$$

**3.** $\dfrac{n}{3n+1}$.

**6.** The problem can be solved in two ways: by induction or by the equation

$$1 - \frac{1}{n^2} = \frac{(n-1)(n+1)}{n^2}.$$

In the solution using induction it is necessary to start with the value $n = 2$.

**8.** $[n(n+1)/2] + 1$. Suppose that $n$ lines have already been drawn. The $(n+1)$st line increases the number of regions by $n+1$.

**9.** Suppose that we can represent $n$ kopeks ($n > 7$) as a sum of three- and five-kopek coins. Show that we can then represent $n+1$ kopeks as a sum of three- and five-kopek pieces. Then apply the method of mathematical induction beginning with $n = 8$.

**14.** Use the inequality

$$\frac{1}{\sqrt{k+1}} > \sqrt{k+1} - \sqrt{k}.$$

**17.** $\Delta u_n = 2n + 1$.

**18.** (a) No; (b) yes. Use the formula $u_1 + \Delta u_1 + \Delta u_2 + \ldots + \Delta u_{n-1} = u_n$.

**20.** The cofficient of the leading term of the polynomial is equal to $k$.

**21.** Use the results of Problems 19 and 20.

**22.** $k!$.

**23.** Solve the problem by induction on $k$. Use the results of Problems 19 and 20.

**25.** Use the results of Problems 19 and 23.

**26.** $n(n+1)(2n+1)/6$. From Problem 25 it is clear that the sum will be expressed by a third-degree polynomial.

**27.** $n(n+1)(n+2)/6$.

**28.** (a) $u_n = u_1 + d(n-1)$; (b) $u_n = u_1 q^{n-1}$.

**29.** (a) $S_n = u_1 n + \dfrac{n(n-1)}{2} d$; (b) $P_n = u_1^n q$ .

**30.** (a) $S_{15} = 35$; (b) $P_{15} = 10^{35}$.

**31.** (a) 0; (b) $4^5 = 1024$.

**32.** For the proof consider the quantity $qS_n - S_n$.

**33.** 210 minutes = 3 hours 30 minutes.

**34.** The horseman spends 0.986 times as much time on the last lap as the cyclist.

**35.** 250,000.

**36.** 164,700. Denote the sum of all positive three-digit numbers by $S^{(1)}$, the sum of all positive three-digit numbers divisible by 2 by $S^{(2)}$, the sum of all positive three-digit numbers divisible by 3 by $S^{(3)}$, and the sum of all positive three-digit numbers divisible by 6 by $S^{(6)}$. Then the sum $S$, which is to be found, is equal to

$$S = S^{(1)} - S^{(2)} - S^{(3)} + S^{(6)}.$$

**37.** $u_1 = 3, d = 6$.

**38.** Yes, it does exist. If 27 is in the $m$th place, 8 in the $n$th, and 12 in the $p$th, the numbers $m$, $n$, and $p$

satisfy the equality $m - 3p + 2n = 0$.

**39.** No, it does not exist.

**40.** The ratio of the progression can assume the values $1, 4, 4 + 2\sqrt{3}$, and $4 - 2\sqrt{3}$.

**41.** Any progression with ratio $(1 + \sqrt{5})/2$ or $(1 - \sqrt{5})/2$.

**42.** The third term of the sequence is also equal to 0.

**43.** $\frac{1}{\sqrt{5}}\left[\left(\frac{1+\sqrt{5}}{2}\right)^n - \left(\frac{1-\sqrt{5}}{2}\right)^n\right]$.

## Chapter 2

**44.** There are various solutions of the problem.

**First Solution.** One can simply add up all of the ways of lighting the kitchen.

(0) There are no lamps lit ●●●●●.

(one way)

(1) One lamp is lit (see the diagram)

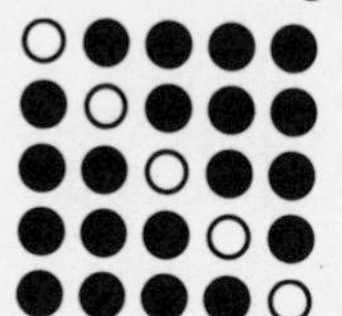

(five ways)

(2) Two lamps are lit (draw a diagram and find the number of ways by yourself).

(? ways)

(3) Three lamps are lit. (Do not hurry. There is no need to draw and calculate anything. Think and you will realize that you already know the answer.)

(? ways)

(4) Four lamps are lit. (It is clear that there are as many ways of lighting the kitchen as there are in Case 1.) Clear?

(five ways)

(5) All five lamps are lit ○○○○○.

(one way)

*Answer*: 32 ways in all

**Second Solution.** It is possible to argue in another way. Suppose that there are not five lamps but one. How many ways of lighting the kitchen are there then? Now suppose that there are two lamps, and so on. By how many times does the number of ways of lighting the kitchen increase when one lamp is added?

**45.** Read the hints for Problem 44.

The use of the symbol $C_n^k$ allows one to substitute the formula $C_5^3 = 10$ for the sentence "In a room there are 5 lamps; there are 10 ways of lighting the room with 3 of the lamps burning." What does the formula $C_5^1 = C_5^4$ denote? What is $C_5^0$ equal to? Use of both methods of solution of Problem 44 leads to the formula $C_5^0 + C_5^1 + C_{5n}^2 + C_{55}^3 + C_{5n}^4 + C_5^5 = 2^5$.

**46.** $3^n$.

**47.** $3^k \cdot 2^{n-k}$.

**48.** Suppose that you have somehow found out how many five-digit numbers there are which do not contain 0 or 8. Can you now compute how many such six-digit numbers there are?

**49.** How many automobile license numbers are there consisting of two digits and one letter? Of two digits and two letters?

**50.** Can you make up a problem about lamps which would be similar to our problem?

**51.** $-5050$.

**52.** If the order of the summands is essential, the number of decompositions is $n - 1$; otherwise, for $n$ even the answer is $n/2$ and for $n$ odd the answer is $(n - 1)/2$.

**53.** $\dfrac{(n-1)(n-2)}{2}$.

**54.** Prove that the sum of the numbers in the $(n + 1)$st row is twice as great as the sum of the numbers in the $n$th row.

**55.** Sometimes the following solution is offered. We arrange 14 bishops as in Figure 43. It is clear that it is impossible to set up more than 14 bishops on the board (so that they do not threaten one another).

Fig. 43

However, as a rule, one cannot explain why this is clear. This solution also fails to shed light upon the second (fundamental) question in the problem.

**56.** $C_5^2 = 10$.

**57. Answer:** $C_{n+k}^k$. (If you are not yet able to write down the formula expressing the answer in terms of $n$ and $k$, we recommend that you content yourself with this answer. An explicit formula will be found somewhat later (see Problems 60, 66, and 67).)

**58.** $C_9^2 \cdot C_7^3$.

**59.** One cannot set up 2 rooks on a single row (so that they do not threaten one another): Thus, at most 1 rook can be placed on each row.

But there are 8 rows and 8 rooks. Thus, exactly 1 rook is in each row. In how many ways is this possible?

**60.** $C_n^2 = \dfrac{n(n-1)}{2}$.

**61.** $C_{28}^2 \cdot (26!) = \dfrac{28!}{2}$ (by $n!$ we mean the product $1 \cdot 2 \cdot 3 \cdot 4 \ldots (n-1) \cdot n$).

**62.** $C_8^5 \cdot 5! = \dfrac{8!}{3!}$.

**64.** $C_8^4 \cdot 8 \cdot 7 \cdot 6 \cdot 5$.

**66.** $C_n^3 = \dfrac{n(n-1)(n-2)}{1 \cdot 2 \cdot 3}$.

**67.** $C_n^k = \dfrac{n(n-1)(n-2)\ldots(n-k+1)}{1 \cdot 2 \cdot 3 \cdot \ldots \cdot k}$. Try to prove this formula by the method of Problem 66.

**68.** $P_{88}^6 = \dfrac{88!}{6!}$ sequences, $C_{88}^6 = \dfrac{88!}{6!\,82!}$ chords.

**69.** In the lines with numbers $2^n$, $n = 1, 2, 3, \ldots$.

**70.** $2^7$.

**71.** $C_7^3 = \dfrac{7 \cdot 6 \cdot 5}{1 \cdot 2 \cdot 3}$.

**72.** $3^{20}$.

**74.** The coefficient of $x^8$ is equal to $C_{56}^8$, the coefficient of $x^{48}$ is equal to $C_{56}^{48} = C_{56}^8$.

**77.** *Hint.* $(a + b)^n = a^n\left(1 + \dfrac{b}{a}\right)^n$.

**79.** First find all terms containing $x^k$.

**80.** Solve a similar problem: Prove that

$$2^n = C_n^0 + C_n^1 + C_n^2 + \ldots + C_n^k + \ldots + C_n^{n-1} + C_n^n.$$

True, this problem has already been met, but now we solve it by another method: In the identity $(1 + x)^n = C_n^0 + C_n^1 x + C_n^2 x^2 + \ldots + C_n^k x^k + \ldots + C_n^{n-1} x^{n-1} + C_n^n x^n$ we set $x = 1$. We get $2^n = C_n^0 + C_n^1 \cdot 1 + C_n^2 \cdot 1^2 + \ldots + C_n^n \cdot 1^n$, as we wanted to prove.

**82.** Three.

**83.** Represent 99 and 101 in the form: $99 = 100 - 1$ and $101 = 100 + 1$.

**84.** $2^5$.

**85.** **Hint:** $10! = 2^8 \cdot 3^4 \cdot 5^2 \cdot 7^1$; can you say how many divisors the number $2^8 \cdot 3^4$ has?

**87.** What can the last digit of our number be? The first digit? **Answer:** 42.

**88.** $2^9$.

**90.** $C_{3n}^n \cdot C_{2n}^n = \dfrac{(3n)!}{(n!)^3}$.

**91.** There are 11 people in the room and only one knows only English.

**93.** One can use the result of Problem 92 or try to count the numbers that do not contain the combination 12. **Answers:** 49401.

## Chapter 3

**95.** The case $+ -$ is impossible. The remaining cases are possible.

**96.** The entries $+ + +$, $- + +$, $- + -$, $- - -$ are possible.

**97.** (a) The shortest of the tall people is always taller than the tallest of the short people. (b) It is changed. For this arrangement both cases are possible.

**98.** A test is said to be difficult if at least one student at each desk failed to solve at least one problem.

**99.** It can.

**100.** Theorems 1 and 6 are true, the rest false.

**101.** Theorem 8 is in the first group; Theorems 2, 3, 6, and 7 are in group 2; and Theorems 4 and 5 are in group 3.

**102.** (a), (b) Consider separately the cases $x > 0$ and $x < 0$. (c) Consider separately the cases $x \leq -\frac{1}{2}$, $\frac{1}{2} \leq x \leq \frac{1}{2}$, and $x \geq \frac{1}{2}$. **Answers:** (a) $x_1 = 1, x_2 = \quad 3$; (b) $x_1 = 1, x_2 = -1$; (c) the segment $[-\frac{1}{2}, \frac{1}{2}]$.

**103.** Consider separately the various arrangements of the points $x$, $y$, and 0 on the number axis. In case (a) equality is possible only when $x \geq 0, y \geq 0$ or $x \leq 0, y \leq 0$. In case (b) it is possible when $x \geq y \geq 0$ or $x \leq y \leq 0$. In case (c) it is possible when $x \geq 0, y \geq 0$ or $x \leq 0, y \leq 0$.

**104.** Assertions (a), (b), and (c) are true; (d) is false.

**105.** True.

**106.** True.

**107.** (b) See Problem 108.

**108.** For sequence (a) all the segments shown are traps. For sequence (b) segments (A) and (B) are troughs, and segment (C) is a trap.

For sequence (c) all three segments are troughs.

**109.** (a) It exists; (b) it does not exist.

**110.** (a) It does not exist; (b) it cannot be determined.

**111.** (a) It exists; (b) it exists.

**112.** The corresponding values of $k$ are shown in the table

| | A | B | C |
|---|---|---|---|
| a | 1 | 1,000 | 1,000,000 |
| b | 0 | $\sqrt{999}$ | $\sqrt{999{,}999}$ |
| c | 0 | $\dfrac{3}{\log 2}$ | $\dfrac{6}{\log 2}$ |
| d | 2 | $2^{1000}$ | $2^{1{,}000{,}000}$ |

**113.** (b) True.

**114.** (b) No one can not. See the sequence of Problem 108c.

**115.** Use Problem 109b.

**116.** (a) The limit is equal to 0; (b) the limit is equal to 1; (c) the limit is equal to 2; (d) a limit does not exist; (e) the limit is equal to 1; (f) the limit is equal to 0; (g) the limit is equal to $\frac{2}{9}$; (h) a limit does not exist; (i) the limit is equal to 0; (j) a limit does not exist.

**117.** No. (**Hint:** See Problem 115.)

**118. Hint:** for (b): Formulate what it means for the number $a$ not to be a limit point of a sequence.

**119.** See Problems 113a, 107a, and 118b.

**120.** (a) 1; (b) 1, $-1$; (c) 0; $\pm\sin 1°$, $\pm\sin 2°$, ..., $\pm\sin 89°$, $\pm 1$; (d) 0; (e) there are no limit points; (f) the limit points are the points of the segment $[0, 1]$.

**121.** The sequence $\{x_n\}$ is unbounded if for each $C$ there exists an $n$ such that the inequality $|x_n| > C$ is satisfied.

**122.** False. See Problem 108b.

**123.** (a) $x_n \to \infty$; (b) $x_n \to \infty$; (c) $\{x_n\}$ is unbounded; (d) $x_n \to \infty$; (e) $\{x_n\}$ is bounded.

**124.** (a) $x_n = (-\frac{1}{2})^n$; (b) $x_n = \frac{1}{n}$; (c) $x_n = \dfrac{n-1}{n}$; (d) $x_n = (-1)^n \cdot \dfrac{n-1}{n}$.

**125.** (1) All sequences satisfy the condition. (2) Not all of the terms of the sequence are equal to $a$. (3) The sequence is bounded. (4) The point $a$ is not a limit point of the sequence. (5) The sequence does not converge to infinity. (6) The number $a$ is not the limit of the sequence. (7) The sequence is bounded. (9) The number $a$ is either one of the terms of the sequence $\{x_n\}$ or a limit point of this sequence.[1]

The remaining conditions are the negations of those

[1]If Condition 9 is satisfied, the point $a$ is called a *closure* point of the sequence $\{x_n\}$.

just listed. That is, condition ($k$) denotes that condition ($17 - k$) is not satisfied. For example, Condition 10 states that Condition 7 is not satisfied, that is, that the sequence is not bounded. Condition 15 states that Condition 2 is not satisfied, that is, that all of the terms of the sequence are equal to $a$; and so on.

**126.** (a)
$$\begin{aligned} &++++-; & x_n &= a;\\ &-+++-; & x_n &= a + \tfrac{1}{n};\\ &--++-; & x_n &= a + 1 + (-1)^n;\\ &--+--; & x_n &= a + n[1 + (-1)^n];\\ &---+-; & x_n &= a + (-1)^n;\\ &----+; & x_n &= n;\\ &-----; & x_n &= a + 1 + \\ & & &\quad + n[1 + (-1)^n]. \end{aligned}$$

**127.** (1) $x_n = \dfrac{1}{n}$; (2) $x_n = \dfrac{n-1}{n}$; (3) $x_n = \left(-\dfrac{1}{2}\right)^n$.

**128.** Prove that if an infinite sequence does not have a greatest term, then it is possible to extract an infinite increasing subsequence.

**129.** Prove that the sequence 1.4, 1.41, 1.414, 1.4142, 1.41422, . . . of approximations of $\sqrt{2}$ is monotone, bounded, and does not have a rational limit.

**130.** Use the result of Problem 128 and the Bolzano-Weierstrass axiom.

**131.** (a) Use the inequality $1/n^2 < 1/n(n-1)$ and prove that the sequence is bounded. (b) Consider separately the subsequences of even-numbered terms and of odd-numbered terms.

**133.** (a) $x_n = \dfrac{1}{n^2}$; $y_n = \dfrac{1}{n}$; (b) $x_n = \dfrac{1}{n}$; $y_n = \dfrac{1}{n}$; (c) $x_n = \dfrac{1}{n}$, $y_n = \dfrac{1}{n^2}$; (d) $x_n = \dfrac{1}{n}$, $y_n = \dfrac{(-1)^n}{n}$.

**134.** (a) $\frac{2}{3}$; (b) 0; (c) 1; (d) 1; (e) $\dfrac{1}{k+1}$.

**135.** Use the identity

$$\sqrt{a} - \sqrt{b} = \frac{a - b}{\sqrt{a} + \sqrt{b}}.$$

**136.** Represent $2^n$ as $(1 + 1)^n$ and use the binomial formula.

**137.** Represent $a^n$ as $[1 + (a - 1)]^n$ and use the binomial formula.

**138.** Use the result of Problem 136.

**139.** 1.

**140.** Take as origin the point where the snail begins. Direct the coordinate axes along the lines of the lattice, and take the side of a square for the unit of measure. Suppose that the first snail is initially at the point $(a, b)$. Find the coordinates $(a_n, b_n)$ of the point where the snail will be after $n$ steps. The slope of the line along which the telescope is directed is equal to $k_n = b_n/a_n$. Prove that $\lim_{n\to\infty} k_n = 1$.

**141.** See the hint for Problem 140. **Answers:** (a) $\lim_{n\to\infty} k_n = 2$; (b) $\lim_{n\to\infty} k_n = 1$; (c) the sequence $\{k_n\}$ does not have a limit.

**142.** Drop perpendiculars from the points $A_0$ and $A_n$ to the $x$-axis. Let $P_0$ and $P_n$ be the bases of these perpendiculars. Prove that the triangle $M_nA_0P_0$ and $M_nA_nP_n$ are similar and compute from this the length of the segment $M_nP_0$. **Answer:** The point $M$ – the midpoint of the segment $OP_0$.

**143.** On a sheet of paper mark off three points representing the school, the theater, and the skating rink. Take any fourth point and construct the line $A_1A_2A_3A_4 \ldots$, representing the boy's path. Compare the segments $A_1A_4$, $A_2A_5$, $A_3A_6$, $A_4A_7$, and so on.

**144.** The point $M$ divides the segment $M_1M_2$ in the ratio 2:1.

**145.** (a) $S_n = (1 - a^{n+1})/(1 - a)$; for $|a| < 1$, $S = 1/(1 - a)$; for $|a| \geq 1$, the series diverges; (b) $S_n = [a - (n + 1 - na)a^{n+1}]/(1 - a)^2$; for $|a| < 1$, $S = a/(1 - a)^2$; for $|a| \geq 1$, the series diverges. (**Hint:** Represent $S_n$ as the sum of $n$ geometric progressions.) (c) $S_n = 1 - [1/(n + 1)]$, $S = 1$; (d) $S_n = \frac{1}{2}\{\frac{1}{2} - [1/(n + 1)(n + 2)]\}$, $S = \frac{1}{4}$; (e) the series diverges. **Hint:** Prove that $1/(n + 1) + 1/(n + 2) + \ldots + 1/2n > \frac{1}{2}$.)

**146.** It is possible to make the "roof" arbitrarily long. **Hint:** Take the length of the bricks to be 1. Prove by induction that the bricks will not fall if the $(n + 1)$st brick (counting from the top) is shifted relative to the $n$th brick by $1/2n$. After this, use the result of Problem 145e.

**147.** $1 + \sqrt{2}$.

**148.** It suffices to compute $x_2$.